英文
听打实训教程

主　编　郭士香
副主编　冯焕红　李淑艳

暨南大学出版社
JINAN UNIVERSITY PRESS

中国·广州

图书在版编目（CIP）数据

英文听打实训教程/郭士香主编；冯焕红，李淑艳副主编．—广州：暨南大学出版社，2011.8（2018.9 重印）
ISBN 978 - 7 - 81135 - 824 - 7

Ⅰ.①英… Ⅱ.①郭…②冯…③李… Ⅲ.①英文—听打—高等职业—教育—教材 Ⅳ.①TP391.14

中国版本图书馆 CIP 数据核字(2011)第 072085 号

英文听打实训教程
YINGWEN TINGDA SHIXUN JIAOCHENG
主编：郭士香　副主编：冯焕红　李淑艳

出 版 人：	徐义雄
责任编辑：	潘雅琴　徐圆圆
责任校对：	林丽旋
责任印制：	汤慧君　周一丹

出版发行：暨南大学出版社（510630）
电　　话：总编室（8620）85221601
　　　　　营销部（8620）85225284　85228291　85228292（邮购）
传　　真：（8620）85221583（办公室）　85223774（营销部）
网　　址：http://www.jnupress.com
排　　版：广州市天河星辰文化发展部照排中心
印　　刷：虎彩印艺股份有限公司
开　　本：787mm×1092mm　1/16
印　　张：11.25
字　　数：260 千
版　　次：2011 年 8 月第 1 版
印　　次：2018 年 9 月第 4 次
定　　价：29.80 元（附送光盘一张）

（暨大版图书如有印装质量问题，请与出版社总编室联系调换）

前　　言

在信息化迅猛发展的电子商务时代，无纸办公（Paperless Office）是理想的办公环境，它不仅能大大地提高工作效率，还能节约能源，这将是社会发展的必然趋势。结合这一趋势以及我国高职英语专业学生就业岗位群的岗位能力需求，还有我国企事业涉外部门工作人员对英文文书处理能力的要求，我们编写了本教材。

本教材以"提高学生实践能力，培养学生的职业技能"为宗旨，从企业对职业院校商务英语专业学生的实际能力需求出发，确定教材特点：任务导向，突出实训操练，强调针对性训练，体现教学改革。

教材内容分四个模块，共八个单元。

模块一（Module 1）基础训练（Basic Practice）共四个单元，具体包括：键盘技巧及盲打（Keyboard Skills and Blind Typing），词、句听打（Words and Sentences Audio Typing），段落听打（Passage Audio Typing），信息摘录（Information Taking），对话听打（Conversation Audio Typing）。

键盘技巧及盲打（Keyboard Skills and Blind Typing）单元介绍了正确打字姿势，标准键盘指法及盲打技巧，同时设计了一系列的英文打字训练项目，例如：字母抄打，短、长句子抄打，段落抄打，还专门设计了潦草手写体文本的辨认及抄打练习。词、句听打（Words and Sentences Audio Typing）、段落听打（Passage Audio Typing）、对话听打（Conversation Audio Typing）三个单元在文稿内容设计难度和打字速度要求及习题安排上既体现从易到难，词、句、段阶梯式训练效果，又注重循环训练，目的在于提高学习者的听打熟练程度及准确率。

模块二（Module 2）高级进阶训练（Advanced Practice）共三个单元，具体内容包括：备忘录和通知，商务信函，会议纪要。此模块主要介绍常见文函格式、排版要求，根据口述要点撰写成文，根据具体语境设计的对话（两人对话、多人对话）和情景介绍来整理并打出以上备忘录、通知、信函及会议纪要等文本。其中在备忘录和通知单元设计了 Word 文档编排技巧。

模块三（Module 3）项目训练（The Project），以 Judy 女士一天工作为线索，根据工作过程和情境设计了一个贯穿项目，把手写文稿抄打、备忘录、通知、商务信函、会议纪要等听打技能全部设计到 Judy 的一天工作过程中。本模块是对全书内容的综合性训练。

模块四（Module 4）英文听打练习（English Audio Typing Practice）适用于学生自学，题型设计以中等难度的信息摘录（Information Taking）为主。

本教材适用：

　　＊ 中职学校英语专业学生听打实训课程教材；

* 高职院校英语专业学生"同声打字"或"英文听打"课程教材；

* 涉外企事业外事部门员工岗前培训教材；

推荐中、高职院校英语专业学生第三或第四学期开设此课程。

选学内容及推荐学时：

1. 中职学校学生课堂学习：

（1）推荐学时：30 学时

（2）选学内容：

Module 1	Basic Practice
Unit 1	Keyboard Skills and Blind Typing
Unit 2	Words and Sentences Audio Typing
Unit 3	Passage Audio Typing
Unit 4	Information Taking
Unit 5	Conversation Audio Typing

2. 高职院校学生课堂学习：

（1）推荐学时：36 学时

（2）选学内容：

Module 1	Basic Practice
Unit 1	Keyboard Skills and Blind Typing
Unit 2	Words and Sentences Audio Typing
Unit 3	Passage Audio Typing
Unit 4	Information Taking
Unit 5	Conversation Audio Typing
Module 2	Advanced Practice
Unit 6	Memos and Notices
Unit 7	Business Letters
Unit 8	Meeting Minutes
Module 3	The Project

3. 自学：

（1）推荐学时：10 学时

（2）选学内容：

Module 4	English Audio Typing Practice

4. 外事部门员工岗前培训

（1）推荐学时：20 学时

（2）选学内容：

Module 2	Advanced Practice
Unit 6	Memos and Notices
Unit 7	Business Letters

前 言

Unit 8　　　　Meeting Minutes

本教材注重学生个性差异，设计了比较宽泛的教学内容。建议使用者或授课教师根据实际需要选择性地删减部分内容，以达到最佳使用效果。另外，本教材配有标准美式发音多媒体教学光盘，以便使用者做听打练习或自学时使用。

本教材由江门职业技术学院郭士香副教授编写第一、第四模块，冯焕红编写第二、三模块，王媛媛、罗光文、冯晓菲、张时平、李祎、黄玲玲、梁颖琦、谢碧玲等为本教材提供了编写材料，辽宁科技学院外语系的李淑艳老师负责本教材的版面、图片设计等。本教材全部录音由来自美国的 Marian Blair 女士及 Howard 先生完成。此外，一些来自广东省本科、高职、中职院校的教学名师，学科带头人，骨干教师等也为本教材的编写提出了建设性意见。

本书的设计、编写、制作过程是一个探索的过程，加之编者水平有限，疏漏和不妥之处恳请专家和读者不吝指正。

编　者

2011 年 3 月 1 日

Contents

前　言	1
Module 1　Basic Practice	1
Unit 1　Keyboard Skills and Blind Typing	1
Unit 2　Words and Sentences Audio Typing	12
Unit 3　Passage Audio Typing	16
Unit 4　Information Taking	20
Unit 5　Conversation Audio Typing	28
Module 2　Advanced Practice	34
Unit 6　Memos and Notices	34
Unit 7　Business Letters	46
Unit 8　Meeting Minutes	58
Module 3　The Project	63
Module 4　English Audio Typing Practice	67
Copy Version and Key to Each Exercise	99
Unit 2	99
Unit 3	103
Unit 4	107
Unit 5	112
Unit 6	118
Unit 7	123
Unit 8	136
Module 3	143
Module 4	146

Module 1　Basic Practice

Unit 1　Keyboard Skills and Blind Typing

Aims and Expectations 能力目标

通过本单元的教学和训练，学生应具备以下能力：
1. 能用正确的指法打字。
2. 能以一定速度盲打。
3. 能在一定准确率要求下盲打。

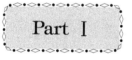

Keyboard and Typing Skills 键盘与打字技巧

1. 正确的打字姿势

正确的打字姿势对于打字的速度和准确率都是非常重要的。如果坐姿不正确，不但会影响打字速度的提高，而且还很容易疲劳、出错。

正确的坐姿应该是：两脚平放，腰部挺直，两臂自然下垂，两肘贴于腋边。手指轻轻接触键盘，手掌离开键盘，不能靠在键盘上。身体可略倾斜，离键盘的距离约为 20～30 厘米。所打字的文稿放在键盘左边，或用专用夹，夹在显示器旁边。打字时眼观文稿，身体不要随文稿倾斜。

2. 标准键盘指法

在键盘的中央有八个手指定位的基本键，分别是"A、S、D、F、J、K、L、;"键，其中"F"和"J"键上都有一个突出的小棱杠，以便于盲打时手指能通过触觉定位。

(1) 双手在基本键上的正确位置应该是：左手小指、无名指、中指和食指应依次虚放在"A、S、D、F"键上，右手的食指、中指、无名指和小指应依次虚放在"J、K、L、;"键上，两个大拇指则虚放在空格键上。在打字时，双手应放在基本键上，当要击打其他键时，手指应从基本键出发，完成击打后，手指必须马上回到基本键来。

(2) 其他键的手指分工：掌握了基本键及其指法，就可以进一步掌握打字键区的其他键位了，左手食指负责的键位有"4、5、R、T、F、G、V、B"共八个键，中指负责"3、E、D、C"共四个键，无名指负责"2、W、S、X"键，小指负责"1、Q、A、Z"及其左边的所有键位。右手食指负责"6、7、Y、U、H、J、N、M"八个键,中指负责"8、I、K、,"四个键,无名指负责"9、O、L、。"四个键,小指负责"0、P、;、/"及其右边的所有键位。这么一划分,整个键盘的手指分工就一清二楚了,击打任何键,只需把手指从基本键位移到相应的键上,正确按键后,再返回基本键位即可。

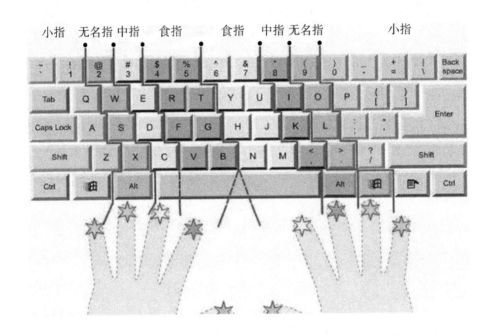

3. 盲打技巧

(1) 盲打的含义

所谓盲打就是在打字时既不看键盘也不看屏幕,只看录入文件的一种打字方式。

在盲打过程中,按照手指分工和标准指法打字,眼和手互不干扰,各负其责,所以打字速度快。熟练掌握盲打的技巧,能大大地提高工作效率。

(2) 盲打的技巧

首先,必须要用正确的指法来打字。由于在打字过程中不允许看键盘,要通过触摸来击打正确的键,因此,指法必须正确,否则准确率就无法保证。各个手指分工明确,各司其职,不要越权代劳,一旦敲错了键,或是用错了手指,可用右手小指击打退格键,重新输入正确的字符。

其次,要真正做到盲打,必须不看键盘,也不看电脑屏幕,只看录入文件。在打字过程中,如果不能专注地看着录入文件,而是看一下键盘,看一下屏幕,再看一下文件,会很容易造成文字的丢失,要重新定位正在输入的文字的位置,不但花了时间,还很容易发生错误。初学者因记不住键位,往往忍不住要看着键盘打字,一定要避免这种情况,实在记不起,可先看一下,然后移开视线,再按指法要求键入。只有这样,才能逐渐做到凭手感而不是凭记忆去体会每一个键的准确位置。

最后,任何技巧都是通过多次反复训练而获得的,因此,初学者只要多记、多练,一

Module 1 Basic Practice

定能达到盲打要求的速度和准确率。

Practice Exercise 实训练习

说明：以下练习题每题有 100~120 个单词，要求在 6 分钟内录入完毕（即 20wpm，每分钟 20 个单词），准确率要求达到 90% 以上（即允许出现的错误不能超过 12 个）。

Task

1. Practice typing letters A S D F J K L and measure the time yourself.（练习打字母 A S D F J K L 并计时）

asdfjkl; asdfjkl; asdfjkl; asdfjkl; asdfjkl;
asdf asdf asdf asdf jkl; jkl; jkl; jkl;
kk jj ss dd ff ll aa ;; jj kk ll ss ff dd aa ;;
ak aj ad ls lf ld kd kl k; ds sd a; sj
ljk lf jkdf fdk fl; askfd fds sdf kl; lkj
; a ; s ; f ; d ; l ; j fj fk fa fs fl f; jf jd js jl ja j;
dj dl dk ds df da d; sj sl sk s; lf ld lj lk l;
ask a dad; as a a sad lad; lad; all fall;
fad ad all lad lass all fall ask dad ads asks ad
sad dad lad fad lad sad dad fall all dads fads

2. Practice typing letters G H V B and measure the time yourself.（练习打字母 G H V B 并计时）

ghvb ghvb ghvb ghvb ghvb ghvb ghvb
hbv hgv vgh ggg hhh vvv bbb gvg hbg
go get lag jag hag gas leg glad sag leg keg egg
she dash had has he lash heed heel hall shell
five cave ever van vac give gave value gave hive
book rob bib bit bob bid rib bad sob bow
fang gang sang hang gang hang fang sang
they are very happy after they heard the good news
it was a very difficult quiz so he failed
beat bowl bad brim better beam bell ball
have above heave brave cab bet crab hell high
ghost good sag lag boy brother baby very good
she left she needing at the sale the things

3. Practice typing letters M N and measure the time yourself. （练习打字母 M N 并计时）

mm nn mn nm mom none more no nothing met
mother month mat jam time mud mam mow mad
same mail most comb meet team man con brim
fan land nat den fan hand sand tan hen than
shine kind gain king hint link sing then then
thin think than keen night knife need fine fang
clam much mince crime cram can munch now
time finish men hamburgers many monkey mice
north northeast noon afternoon kneeling sang
won want sun fun Westminster went Nancy nana
Mary did not bring his material to the meeting.
He is a fine dad; she says he needs his kind

4. Practice typing letters O R and measure the time yourself. （练习打字母 O R 并计时）

fro fro or ro oor roo or ro dor dro mor
rag her rat ran far free jar dart tar the dirt fir
lol lot log hold done too to go do soon old
tooth good song jolts fool sold told gold fold
aLa aJa aKa aHa aHa aOa aOa aNa aNa
done fore note nose none sore soar dare tore lone fir
for this; for her; for their; for those; for him; tar for the jar
we asked Jane to send the pizza to Mary;
there are the letter that she sent to Nancy;
is to to the go for decker or and she that this these

5. Practice typing letters E T and measure the time yourself. （练习打字母 E T 并计时）

he see tee fee kee led feed sea fed dead seas
steel lead task these dash teeth eat feet
she dash had has he lash heed heel hall
jet fat tea set let sat tall talk tell eat
he has a deal; dad had the jet; ; the meal;
deal the deal; at least love a mom; a fast cat;
shell sheds hash heal ashes ash
fat east sat eat jets feat teak least let
Let's go to the cinema to see a film together.
Cathy prefers coffee to tea, while Kate prefers tea to coffee.
He has to take bags when he goes out for camping.
They were very happy after they heard the good news.

Module 1 Basic Practice

6. Practice typing letters X Y Z and measure the time yourself.（练习打字母 X Y Z 并计时）

hex fax fix six box flex next fox ox

wex box hex axe six jinx fox tax wax

your yes joy jay yoyo you yet year yum

year you yard yes jay ray yam say yawn

zip zone zoom zest zap zany lazy zero

zig dozen zag zip zinc zone graze faze

azav azav azav azav azav azav azav azav

sly why day fly try stay bay way pay may

doze frizz leave quiz froze freeze gaze size zoo maze gaze

She bought six boxes in the yard.

They were not sure whether they have time to go to the zoo.

You should finish your assignment by the end of this week.

7. Practice typing letters U W and measure the time yourself.（练习打字母 U W 并计时）

sws sws sws sws sws sws sws sws sws sws

juj juj juj juj juj juj juj juj juj juj juj juj juj

jug run jut just dug hug rug our use sun fun

wet were sew saw sow wig was won win we

few were we sat was wag drag dare wade date

junk fuss our use us four work down town two

week won while with will want wall would well

wig were few sat was wag wade wet rude waste

Let's go. Ned wants to buy a white cup.

I was drunk last night.

I saw a wire in the street.

I came. I saw. I conquered.

Nancy went to Westminster.

Kathy went to the store.

8. Practice typing letters I C and measure the time yourself.（练习打字母 I C 并计时）

fist kid kit hid hill sit fit it fill sill little

shine kind gain king hint link sing interest

cup cut cow cod can tack call cot cat cell ceiling

clam cost cab cob crib cast computer con car

if in it a it is in is it knife big needing something

time is important think this thin hail tail fail laid nail

camera catch cough caught cut country come came

crime cram crab mince nice advice practice like Cindy

single sister nothing hit in face his kind instead
height night fight light frighten sign sigh night light
century cent currency current caught cunning case
Can you catch the candy I throw to you?
He is a fine dad and she says she needs his kindness.

9. Practice typing letters P Q and the left and right SHIFT key, and measure the time yourself.（练习打字母 P Q 及左右 SHIFT 键，并计时）

up op ap qp wp ps pp pup quq pup pop qoq oq op
?p? ?p? ?p? ?p? ?p? ?p? ?p? ;p; ;p; ;p; ;p; ;p; ;p; ;p;
aQa aQa aQa aQa aQa aQa aQa aQa aQa aQa
pep pet pen pin part pug pan pup pop pat pot lap pal
past pest president spare hip ship part-time please
question quit quiet queen quite quip quirt equip
quote quits quilt squall quench squid quick quip queen
To Paul, Ed, Sal, Fred, Don, Dan, Sal, Dean, Gorge
When Where Who Whom What Want Waste
Opportunity Opinion Onion Oh Only Oxygen Ox
She is sending the gift to Peter.
I was not quick enough to quote all this.

10. Practice typing numbers and measure the time yourself.（练习打数字并计时）

1 2 3 4 5 6 7 8 9 0 9 8 4 2 6 7 5 0
23 45 66 81 49 26 52 90 75 63 69 18 97
189 264 321 840 439 744 539 105 372 844
3,429 4,874 5,309 4,666 9,065 7,583 8,210 2,351
44,890 32,819 51,800 43,631 52,504 79,310 94,524
321,095 540,000 670,333 742,500 961,728 253,749
8,655,712 9,499,356 8,254,719 4,573,800 7,436,255
57,345,820 61,822,543 20,540,000 79,734,462 85,641,300
320,563,891 904,583,472 441,309,582 288,357,911
1,540,916,758 5,622,504,302 7,280,122,524 1,010,051,117

11. Type the following sentences and measure the time yourself.（抄打下列句子并计时）

1. Practice makes perfect and constant practice is the only way to achieve becoming a good typist.
2. The way to greet a British person is to shake hands with him or her.
3. Kissing is not common except in the case of elderly ladies.
4. I enjoy playing computer games a lot.
5. I think American football is very exciting, but I don't like playing basketball.
6. Many of my classmates are working as volunteers.
7. The Department of Labor did a study on how Americans use their free time.

Module 1 Basic Practice

8. The advantage of green food in the market is that this kind of food can bring no harm to people's health.

12. Type the following sentences and measure the time yourself. （抄打下列句子并计时）

1. Now, the green food in China is graded into two levels, that is, A level and AA level.
2. In 1950, Diners Club and American Express launched their charge cards in the USA, the first "plastic money".
3. Did you know that taking your medications properly is one of the best ways to avoid future health care costs?
4. The transportation in Beijing has changed a lot and there are more subways and light rails.
5. I have published two short stories in national magazines, and I have also finished my novel.
6. Most people hate the terrible morning sound of an alarm clock when a loud noise wakes you up from your sweetest dreams.

13. Type the following sentences and measure the time yourself. （抄打下列句子并计时）

1. When I first came to New York and studied in a university in 1997, everything seemed new and strange.
2. Bus is one of the most convenient means of transportation and some of them have comfortable seats and toilets.
3. The main obvious purpose of advertising is to inform people of the available products or services.
4. A successful business trip highly depends on a detailed and thoughtful schedule.
5. As a general rule, the businessperson should keep the schedule flexible enough to allow for both unexpected problems and opportunities.
6. From the beginning of time, people have wanted to get from one place to another.

14. Type the following sentences and measure the time yourself. （抄打下列句子并计时）

1. You can do a lot during a working holiday—picking apples on a farm or teaching English in Thailand.
2. William has asked me for a loan of five pounds and I don't know whether I did right by lending him the money.
3. The whole class is divided into three big groups: the Land Transportation Group, the Water Transportation Group and the Air Transportation Group.
4. These books cover every area of life, from love to work and back again.
5. These "feminine" books ask you to search deep inside you for the answers to life's problems.
6. They offer rules you can use to change the state of your heart and everything else follows.

15. Type the following sentences and measure the time yourself. （抄打下列句子并计时）

1. Piggy Wiggy was the first self-service store which was opened in 1916. Prior to 1916, when Clarence Saunders invented the self-service food store and named it Piggy Wiggy, shopping for food was an entirely different experience from what it is today.

2. In the past, department stores were the only modern trade that was known among Chinese shoppers, but ever since China's open door policy and China's rapid transformation from a planned economy to a market economy, great changes have taken place in the retail market.

3. Wal-Mart, the world's largest retailer, was founded in America forty years ago. The company is a giant in every respect, ranking first on the Fortune 500 list.

16. Type the following sentences and measure the time yourself.（抄打下列句子并计时）

1. The weekend was over, and Begbie had returned to town, restless, and strangely unhappy. There was within him a curious sense of something lost. He packed his suitcase silently but could find nothing missing.

2. Upon my return from the never-to-forgotten series of golden hours at Sea Cliff, after the habit of the departing guest, I have at least left one of my possessions behind me.

3. If by some good chance you have found it, and it's useless to you, will you be good enough to return it to me? Or, if by some good fortune you find it worth retaining, will you please tell me it is in your custody and is not lying somewhere neglected? It is the only one I have, and it has never passed out of my keeping before.

17. Type the following paragraphs and measure the time yourself.（抄打下列段落并计时）

1. This is the story of how my husband got mugged at the chemist's. It wasn't his fault. On certain matters, he is as innocent as a newborn baby and should not be allowed to roam around unsupervised at cosmetics counters. In other words, he is a genuine guy. Genuine guys are sometimes known as retro-sexual; to distinguish them from metro-sexual, who are men with the good taste of a woman.

2. Secretly, I've always thought my husband could stand to be just a little bit more metro. Sometimes I buy him fancy shaving cream and leave it suggestively on his side of sink. He never gets the hint. He prefers a ten-second dry shave with a plastic disposable razor and toilet paper to stanch the wounds.

18. Type the following paragraphs and measure the time yourself.（抄打下列段落并计时）

1. ISO is a worldwide federation of national standards bodies from more than 140 countries, one from each country. ISO is a non-governmental organization established in 1947. The mission of ISO is to promote the development of standardization. ISO's work results in international agreements which are published as International Standard.

2. The art of bargaining is a very specialized art. It is the art of getting the most and paying the least. Shop around a little first to develop a sense of what things should cost. Let the seller make the first offer, then counter with a reasonable offer somewhat less than you expect to pay. Once you've decided on an object, let your attention wander so that the seller doesn't know you're hooked.

19. Type the following passages and measure the time yourself.（抄打下列短文并计时）

1. Couples usually ask for advice when they are just about ready to throw in the towel. Their Love Banks have been losing Love Units so long that they are now deeply in the red. And

Module 1　Basic Practice

their negative Love Bank accounts make them feel very uncomfortable just being in the same room with each other. To be in love again means they must re-deposit all the Love Units that were withdrawn. In order to deposit enough Love Units to all in love, they must follow rules that they don't feel, like the rule of care, the rule of protection, the rule of honesty, and the rule of time.

2. We live in a global village, but how well do we know and understand the cross-cultural influence on business? Maybe we assume that the widespread understanding of the English language means a corresponding understanding of English customs. However, it might not be necessarily the case. If we want to have good business, we must understand different cultures in doing business. Before you go to a country on business, you should have to learn something about her culture. When in Rome, do as the Romans do.

20. Type the following dialogue and measure the time yourself.（抄打下列对话并计时）

Waiter: Good evening, sir. Here is the menu for you.

Johnson: Thank you.

Waiter: May I take your order, sir?

Johnson: Sure. I'd like to try some Sichuan food today. Would you recommend some?

Waiter: May I suggest Kung-pao chicken? It's typical Sichuan dish, spicy and hot.

Johnson: OK, I'll take it.

Waiter: I would also suggest Bean-curd with pepper and chili sauce. Many customers like it.

Johnson: It sounds interesting. I'll have a try.

Waiter: What kind of drink do you prefer, beer, wine or tea?

Johnson: I'd like to try some Chinese tea. What kind of tea would you suggest?

Waiter: Chinese green tea is always a favourite with our customers.

Johnson: OK, I'll have some Chinese green tea, then.

Waiter: Thank you. Your ordered food will be ready in a few minutes.

21. Type the following dialogue and measure the time yourself.（抄打下列对话并计时）

Manager: Hello, I'm the service manager. Is there anything I can do for you?

Johnson: I want to complain about the clerk. It's unfair.

Manager: Can you tell me what happened?

Johnson: Of course. I've been waiting in a queue for a long time. But, unexpectedly, he first served that man over there. How can things be like this?

Manager: If that's so, it's certainly his fault. We apologize to you in advance, and we hope you'll forgive us. Now I'll ask him to apologize to you. Is that OK?

Clerk: I'm sorry to have kept you waiting so long. That customer arrived here earlier and then he went away for an emergency. He came here again to continue the service. It's my fault not to have explained it to you.

Johnson: Oh, it doesn't matter. I don't care now that you have given me a clear explanation.

Manager: From now on, we must improve our service to satisfy our customers.

Johnson: That sounds better.

Manager: We sincerely accept your suggestions.

22. Type the following hand-written material according to the requirement and time yourself. (抄打下列手写材料并计时)

Cathy, please type this letter for me and send it to the following address. Thank you!
Cecelia

January 24, 2010
Pacific Co. Ltd.
Room 303, Golden Commercial Building
711-715 Container Port Road
Kwai Chung

Dear Sirs,
Subject: ORDER NO. 45E

We thank you for your letter dated January 20, in which you placed a further order for 8 dozen shirts / ladies'.

We have been forced to increase the price by 5%, effective immediately, of most of our goods and our customers were informed by circular letter. However, we have discovered that your company was not included in this notification through an oversight for which we take full responsibility!

To show our appreciation of you as a valued customer, over many years, we will complete this order at the original price agreed on before for these garments.

Yours sincerely,
Cecilia Man (Mrs)
Sales Manager

Please find enclosed a copy of our catalog with the revised prices.

Module 1 Basic Practice

23. Type the following hand-written material according to the requirement and time yourself. （抄打下列手写材料并计时）

Memo

To: Peter Wang, Purchasing Manager
From: Clarence Kan, Senior Purchasing Clerk
Date: 28 February, 2010
Subject: Man Ho Stationary Co. Account

I have looked into the matter of the letter sent by Man Ho Stationary. According to my records, they were wrong to demand payment of $8,750.00 because their statement of account doesn't show two things:

1. On Feb 16, a cheque for $56,900 was paid into their account.

2. They mistakenly sent us 100 reams of typing paper instead of the 75 reams ordered.

If these two factors are taken into account, we do not owe them anything as I sent them a cheque for $1,850 at the end of last week (see copy attached) to clear the account.

Simon Kan

Enclosed: as stated in memo.

Unit 2　Words and Sentences Audio Typing

Aims and Expectations 能力目标

通过本单元的教学和训练，学生应具备以下能力：
1. 能听打单词。
2. 能听打句子。

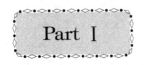

Audio Typing Skills 听打技巧

1. 听打的含义

所谓听打就是边听英文录音材料边把它输入电脑。录音材料的播放速度为 60 字/分钟，要求学生具备将所听录音迅速整理成电子稿的能力。它是一种双手、耳朵、大脑的协调运动，耳朵听到信息传输给大脑，大脑指挥手指打字，不能只顾着听忘了打，也不能只顾着打而忘了听。

2. 听打技巧训练

（1）听打的基本技巧——连打（Continuous Typing）

听打材料的语速为 60 字/分钟，比正常人说话的语速稍慢，但即便如此，也远比我们听到材料后再输入电脑的速度快，因此，我们很难做到打字速度与录音播放速度同步，只能一边听，一边记忆，一边打字。要想把所有的材料记住似乎也很难做到。因此就要采用"连打"的技巧。

所谓连打，就是在听录音材料时，先听开头的几个字，记下来，然后才开始打，在打这几个字的同时，耳朵又继续听接下来的几个字，再记下来，以此类推。这种打字技巧要求我们在听录音的时候，不是从听到第一个字就开始打，也不是听完整个句子才开始打，而是听 3～4 个字，在整个听打的过程中，我们的脑海里应该储存着接着要打的几个字。

（2）数字的听打

数字一般采用阿拉伯数字来打较为简便，例如单独的数字、时间、日期、尺寸和金额等。而当数字"1"单独出现时，要打完整的单词"one"；当数字是句子的开头时，则要打完整的单词。

（3）标点符号的听打

在本教材的听打训练中，标点符号在录音中也读出，要求听打下来。见下表：

Module 1 Basic Practice

comma，逗号	full stop．句号	colon：冒号
semicolon；分号	hyphen - 连接号	dash—破折号
question mark？问号	open parentheses（ 左括弧	close parentheses）右括弧
open quotation mark " 左引号	close quotation mark " 右引号	center 居中
bold 加粗	underline 下划线	italicize 斜体
paragraph 另起一段	initial capital 首字母大写	all capitals 所有字母大写

Practice Exercise 实训练习

Task

1. Listen and type the punctuation marks and numbers according to the recording.（听打标点符号及数字）

Exercise 1

In this exercise，you will hear the punctuation marks that you need to include in your typing.（请把听到的标点符号打出来）

Exercise 2

In this exercise，you will hear some numbers and you are required to type them in complete WORDS.（请把听到的数字用英语单词打出来）

Exercise 3

In this exercise，you will hear some numbers and you are required to type them in FIGURES.（请把听到的英语数字用数字符号打出来）

2. Listen and type the words according to the recording. Pay attention to their spellings and meanings.（根据录音听打，注意单词拼写及意义）

Exercise 1

In this exercise，you will hear 30 words and you are required to type them. Each word will be read three times and there is a pause between every word.（听打30个单词。每个单词读三遍，每两个单词之间，有一个短暂的停顿）

Exercise 2

In this exercise，you will hear 30 words and you are required to type them. Each word will be read twice and there is a pause between every word.（听打30个单词。每个单词读两遍，每两个单词之间，将有一个短暂的停顿）

英语听打实训教程

Exercise 3

In this exercise, you will hear 30 words and you are required to type them. Each word will be read only once and there is a pause between every word. （听打30个单词。每个单词只读一遍，每两个单词之间，有一个短暂的停顿）

3. Listen and type some short sentences according to the recording which is read at 60 words per minute. Pay attention to the punctuation direction "full stop". （听打下列短句，语速为60字/分钟。注意"句号"标点符号）

Exercise 1

Listen to the short sentences the first time without typing and try to remember them. Then, listen for the second time and type the sentences. If necessary, listen for the third time, complete the sentences or correct your mistakes. （听打下列短句，每句听三遍。第一遍，只听不打，尽量记住句子的意思。第二遍，听录音并打出句子。第三遍，把句子补充完整或改错）

Exercise 2

Listen to the short sentences the first time without typing and try to remember them. Then, listen for the second time and type the sentences. If necessary, listen for the third time, complete the sentences or correct your mistakes. （听打下列短句，每句听三遍。第一遍，只听不打，尽量记住句子的意思。第二遍，听录音并打出句子。第三遍，把句子补充完整或改错）

Exercise 3

Listen to the short sentences the first time without typing and try to remember them. Then, listen for the second time and type the sentences. If necessary, listen for the third time, complete the sentences or correct your mistakes. （听打下列短句，每句听三遍。第一遍，只听不打，尽量记住句子的意思。第二遍，听录音并打出句子。第三遍，把句子补充完整或改错）

4. Listen and type some longer sentences according to the recording which is read at 60 words per minute and practice the skills of continuous typing. Pay attention to the punctuations. （听打下列长句，练习连打技巧，语速为60字/分钟。注意句子中的标点符号）

Exercise 1

Listen to the following longer sentences, not the whole sentence, but the first few words before beginning to type and then listen to the next few words while typing. You should always have some words in mind while listening to the next ones until you finish typing the whole sentence. You can listen to the sentences three times. （听打下列长句，运用连打技巧，听开头几个单词后开始打，而不是听完整个句子才开始。把前面的几个单词记住，并一边打，一边听接下来的几个单词，以此类推，直到打完整个句子为止。听三次录音）

Exercise 2

Listen to the following longer sentences, not the whole sentence, but the first few words before beginning to type and then listen to the next few words while typing. You should always have some words in mind while listening to the next ones until you finish typing the whole

Module 1 Basic Practice

sentence. You can listen to the sentences three times. (听打下列长句，运用连打技巧，听开头几个单词后开始打，而不是听完整个句子才开始。把前面的几个单词记住，并一边打，一边听接下来的几个单词，以此类推，直到打完整个句子为止。听三次录音)

Exercise 3

Listen to the following longer sentences, not the whole sentence, but the first few words before beginning to type and then listen to the next few words while typing. You should always have some words in mind while listening to the next ones until you finish typing the whole sentence. You can listen to the sentences three times. (听打下列长句，运用连打技巧，听开头几个单词后开始打，而不是听完整个句子才开始。把前面的几个单词记住，并一边打，一边听接下来的几个单词，以此类推，直到打完整个句子为止。听三次录音)

Test

This part is to test your audio typing skills. You are required to listen and type the following sentences. Assess yourself by the criteria followed. (测试你的听打能力。听打句子。用后面的评分标准表给自己评分)

等级	评分标准
优（90~100分）	1. 指法正确无误； 2. 准确率达到90%以上（错误不超过14个）。
良（80~89分）	1. 指法错误不超过2个； 2. 准确率达到85%以上（错误不超过21个）。
中（70~79分）	1. 指法错误不超过3个； 2. 准确率达到80%以上（错误不超过28个）。
及格（60~69分）	1. 指法错误不超过4个； 2. 准确率达到75%以上（错误不超过35个）。
不及格（60分以下）	1. 超过4个指法错误； 2. 准确率低于75%（错误超过35个）。

Unit 3　Passage Audio Typing

Aims and Expectations 能力目标

通过本单元的教学和训练，学生应具备以下能力：
1. 听打段落。
2. 听打短文。

Audio Typing Skills 听打技巧

在听打段落和短文时，由于内容较多，精神必须高度集中，听取每一个单词，记录下来并打入电脑。稍不留神，信息就会丢失，导致文段无法完成。因此应注意以下几点：

1. 长单词、专业名词的听打技巧

在听打大段文字时，我们需要运用速记的技巧，特别是其中包括一些较长的单词和专业名词在听打中耗时多，往往容易因为打这些单词而错过后面的内容。所以当遇到此类单词和专业名词时，我们一般使用听力的速记法，只把开头的音节先打出来，或者使用它的缩略词。例如：

经济：Eco　　教育：Edu　　文化：Cul　　政治：Po　　科技：Sc
旅游：Tra　　环境：Env　　工业：Ind　　农业：Agr
As soon as possible（尽快）：ASAP　　Look forward to（期待）：LFT

2. 错误发生时的处理方法

在输入大量文字材料时，录入错误是无法避免的。所以在听打训练中，有错误发生时，应保持冷静，不要急着去改，而应该马上在旁边键入正确的字母或单词，以便记忆。等整个文段输入完毕后，再回头修改。

3. 遇到听不懂的信息时的应对方法

听打材料时，有时会遇到个别听不懂的单词或者短语，尽量把它的发音打下来，不要因为听不懂而停下来去想，这样只会让你丢失需听打的更多的信息，得不偿失。

Module 1　Basic Practice

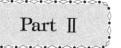

Practice Exercise 实训练习

Task

1. Transcribe four short paragraphs of about 40 words from the recording at 60 words per minute with the technique of continuous typing. （听录音，打出四段约 40 个单词的段落，语速为 60 字/分钟。注意运用连打技巧）

Exercise 1

Clues：

Spring Festival 春节	traditional 传统的	celebrate 庆祝
couplet 对联	lantern 灯笼	dumpling 饺子
relative 亲戚		

Exercise 2

Clues：

throw 扔　　rubbish 垃圾　　responsible 有责任的

Exercise 3

Clues：

symbolic 象征的	meaning 意义	European 欧洲的
prosperity 繁荣，兴旺	imperial 帝王的，皇室的	death 死亡
funeral 葬礼	symbolize 象征着	misfortune 不幸

Exercise 4

Clues：

traveling 旅行	United States 美国
famous 著名的	Central Park 中央公园

2. Transcribe four short paragraphs of about 60 words from the recording at 60 words per minute with the technique of continuous typing. （听录音，打出四段约 60 个单词的段落，语速为 60 字/分钟。注意运用连打技巧）

Exercise 1

Clues：

Thanksgiving Day 感恩节	relate to 与……有关	Christianity 基督教
consider 看作，认为	offer 提供	family gathering 家庭聚会
turkey 火鸡	cranberry sauce 小红莓果酱	parade 游行，巡游

17

Exercise 2
Clues：
math 数学　　　　　　biology 生物学　　　　　　geography 地理学
politics 政治学　　　　physics 物理学

Exercise 3
Clues：
business relation 商务关系　　in compliance with 顺从　　request 要求
separate 独立的，分开的　　　catalogue 产品目录　　　　range 范围
irrevocable 不可撤销的　　　　confirmed 保兑的　　　　　Letter of Credit 信用证

Exercise 4
Clues：
difference 区别　　　　eating habit 饮食习惯　　　　dish 一道菜
host 主人　　　　　　 cuisine 烹饪艺术　　　　　　hospitality 好客

3. Transcribe four short paragraphs of about 80 words from the recording at 60 words per minute with the technique of continuous typing.（听录音，打出四段约 80 个单词的段落，语速为 60 字/分钟。注意运用连打技巧）

Exercise 1
Clues：
Mid-Autumn Day 中秋节　　　moon-cake 月饼
celebrate 庆祝　　　　　　　reunion 团圆

Exercise 2
Clues：
enjoyable 愉快的，快乐的　　grandparent(s) 祖父母　　countryside 乡下
neighbourhood 四邻　　　　　realize 意识到　　　　　　knowledge 知识

Exercise 3
Clues：
impolite 不礼貌的　　　　acceptable 可接受的　　public 公共场合
surprising 令人吃惊的　　nor 也不　　　　　　　necessary 需要的
crowded 拥挤的　　　　　customs 习俗，风俗　　　differ 不同，有差异
uncomfortable 不舒服的

Exercise 4
Clues：
dim sum 点心　　　　　serve 供应　　　　wrap 包、裹
pastry 面皮、面团　　　patron 顾客　　　　provide 提供
chopsticks 筷子

Module 1 Basic Practice

4. Transcribe three short paragraphs of about 100 words from the recording at 60 words per minute with the technique of continuous typing. （听录音，打出三段约100个单词的段落，语速为60字/分钟。注意运用连打技巧）

Exercise 1
Clues：

ideal 理想的	career 职业	chiefly 主要地
freedom 自由	opportunity 机会	acquire 获得
effort 努力	purify 净化	soul 灵魂

Exercise 2
Clues：

enclosure 附件，附寄物	distributor 经销商，分销商	aid 辅助（工具）
booth 摊位	Toy Fair 玩具交易会	Earl's Court 伯爵宫（展厅）
appointment 约定，预约	exhibition 展览	staff 员工

Exercise 3
Clues：

The Sports Meet 运动会	Science Week 科学周	Art Week 艺术周
respect 方面	helpful 愿意帮忙的	society 社会

Test

This part is to test your audio typing skills. You are required to listen and type the following passage of about 100 words. Assess yourself by the criteria followed. （测试你的听打能力。听打出约100个单词的段落。用后面的评分标准表给自己评分）

Clues：

Lantern Festival 元宵节	lunar year 农历	riddle 谜语
eve 前夕，前夜	wonderful 精彩的	folk performance 民谣表演
Yangko 秧歌		

等级	评分标准
优（90~100分）	1. 指法正确无误； 2. 准确率达到90%以上（错误不超过11个）。
良（80~89分）	1. 指法错误不超过2个； 2. 准确率达到85%以上（错误不超过17个）。
中（70~79分）	1. 指法错误不超过3个； 2. 准确率达到80%以上（错误不超过23个）。
及格（60~69分）	1. 指法错误不超过4个； 2. 准确率达到75%以上（错误不超过29个）。
不及格（60分以下）	1. 超过4个指法错误； 2. 准确率低于75%（错误超过29个）。

Unit 4　Information Taking

Aims and Expectations 能力目标

通过本单元的教学和训练，学生应具备以下能力：
能听懂对话，摘录主要信息，并打入电脑。

在实际工作情境中，经常遇到这类记录：草草记下别人对话的关键词，以便后期整理或查询。听打此类对话时，边听边记录主要信息，例如：时间、地点、人物、事件等。要求速打关键内容即可，左侧栏目是句子时，请用句子记录；左侧栏目是短语时，请用短语记录。

Example

有人到北京出差，晚上入住一家酒店，现在在前台投诉。你将听到以下对话，请按要求填充表格：

W：Excuse me.

M：What can I do for you, madam?

W：I'm staying in Room 1605. I want to have a rest now, but the guests next door are making too much noise. I hope you can do something.

M：I'll send a clerk to handle this matter immediately.

W：Thank you. And can you give me some more clean towels?

M：Of course.

Items	Information
Room Number	1605.
Complaint	The guests next door are making too much noise.
Who to handle the problem	A clerk to handle this matter.

Practice Exercise 实训练习

Task

1. Listen to the following dialogues, take down the key information and fill in the forms.（听对话，摘录主要信息，并填表）

注：每题10分，每空2分。

Module 1   Basic Practice

Exercise 1

汤姆和玛丽正在讨论最近的天气。你将听到以下对话，请按要求填充表格：

Items	Information
Weather today	
Weather Mary likes	
What is Tom going to do this weekend?	
Weather in London now	
Season in Sydney now	

Useful Expressions 常用句型

询问天气：

1. What's the weather like in London in this season? 在这个季节伦敦的天气怎么样？
2. How about the weather in Sydney now, Mary? 玛丽，现在悉尼的天气怎样？
3. What's the weather like today? How is the weather today? 今天的天气怎样？
4. What does the weather forecast say? 今天的天气预报怎么说的？
5. What's the temperature today? 今天的气温怎样？

描述天气：

1. It says in the newspaper that the weather today will be rainy and cold again. 报纸说今天的天气有雨，而且寒冷。
2. And sometimes it snows for a few days. 有时会持续好几天下雪。
3. Pretty hot. Now it is summer there. 非常炎热。那里现在是夏天。
4. It looks as if it is going to rain. 看来要下雨了。
5. It's sunny. / It's clearing up. 天晴了。
6. There will be a storm tomorrow. 明天要有一场风暴。

描述个人对天气的喜恶：

1. Well, to tell you the truth, I don't like this kind of weather. 老实告诉你，我不喜欢这样的天气。
2. I prefer it warm and dry. 我喜欢温暖干燥的天气。
3. Weather like this makes me feel tired. 这样的天气使我感到疲倦。
4. A lovely day, isn't it? 天气真好，不是吗？
5. I hope it stays fine. 我希望天气一直晴朗。
6. It's good to see the sun again. 真高兴又看到太阳了。
7. It's rather changeable, isn't it? 天气变化无常，不是吗？

Exercise 2

莎莉邀请李明去她家吃饭，期间他们聊得很开心。你将听到以下对话，请按要求填充表格：

Items	Information
Food Sally likes	
Li Ming's favourite food	
Food Li Ming suggests	
Ingredients in Shanghai food	
Time to taste some new food	

Useful Expressions 常用句型

1. Dinner is ready. 晚餐准备好了。

2. I didn't know you are such a good cook. 我不知道你是那么好的厨师。

3. All the dishes look so delicious. 所有的菜看起来那么美味。

4. Just help yourselves to anything you like. 爱吃什么自己动手。

5. It is a surprise for me that you know how to make Cantonese food. 你会做广东菜真让我吃惊。

6. I like Cantonese food. It is light and tastes good. 我喜欢广东菜。它比较清淡而且美味。

7. But my favourite is Sichuan food. 但是我最喜欢的是四川菜。

8. Sichuan dishes are delicious, but they are too spicy. 四川菜很好吃,但它们太辣了。

9. It is a bit heavier than Cantonese food, and it uses a lot of seafood and fish. 它比广东菜口味重一点,而且用大量的海鲜和鱼来做菜。

10. I'm fond of seafood. 我喜欢吃海鲜。

11. I know a famous restaurant that serves very good Shanghai food. 我知道一家有名的餐馆提供很好吃的上海菜。

Exercise 3

你是一个中等职业学校的应届毕业生,现在一家公司接受面试,你将听到以下对话,请按要求填充表格:

Items	Information
Interviewee's name	
Major	
Language ability	
Computer skill	
Hobbies	

Useful Expressions 常用句型

1. Please tell me something about your education first. 请先告诉我你受教育的情况。

2. My major is Business English. 我的专业是商务英语。

Module 1　Basic Practice

3. What is your language ability? 你的语言能力怎样?
4. I passed PETS – 2. 我通过了公共英语二级。
5. I am familiar with Microsoft Office software. 我能熟练操作微软办公软件。
6. We'll inform you soon. 我们会尽快通知你。

Exercise 4

一位顾客来到商场投诉新买的货物有问题，店员正在跟他讨论解决办法。你将听到以下对话，请按要求填充表格：

Items	Information
What did the man buy?	
What's the problem?	
What did he hear when he switched the MP3 player on?	
How will the woman solve the problem?	
What does the woman ask for?	

Useful Expressions 常用句型

1. May I help you? 有什么事吗?
2. We had a damaged shipment from you. 你们送来的货有损坏。
3. We'll look into it right away for you. 我们会立刻调查清楚。
4. Just whose fault is this damage? 这次损坏究竟是谁的责任?
5. The goods were in good shape when they left our store. 货离开商店时都是完好无缺的啊。
6. It certainly didn't arrive here that way. 送到这儿时可不是那样!
7. Here is the final settlement for your claim. 你的赔偿问题终于解决了。
8. Thanks, we appreciate the fast work. 谢谢你们这么快就办好了。
9. We only hope we won't have this kind of problem again. 我们仅希望不会再有这样的事情发生。

Exercise 5

有人想去颐和园，但是不知道怎么走，在问路，路人给他指了如何去那里。你将听到以下对话，请按要求填充表格：

Items	Information
Destination	
How far is it?	
Which turn?	
To the right or left?	
Is it far?	

Useful Expressions 常用句型

1. No, but I can ask the way. Excuse me, can you tell me where the Prince's Building is? 不认识，但我可以问路。打扰一下，你能告诉我王子大厦在哪里吗？
2. I'm sorry. I'm a stranger here myself. 很抱歉，我也不熟悉这里。
3. Excuse me. How do I get to the Prince's Building, please? 打扰了，请问王子大厦怎么走？
4. Well, turn to the left at the first corner after the crossroads. It's near the corner. You can't go wrong. 噢，过十字路口后在第一个拐角向左拐，离拐角不远。你会找到的。
5. Is it far from here? 离这儿远吗？
6. No, it's only a couple of blocks away. 不远，过两个街区就到了。
7. Don't mention it. 不客气。
8. Can you tell me the way to the bus station? 请问去汽车站怎么走？
9. Go straight ahead and turn left at the traffic lights. 沿这条路一直向前走，在红绿灯那儿向左转。
10. Would you please tell me where the post office is? 请问邮局在哪儿？
11. Would you please tell me if there is a hospital nearby? 请问附近有医院吗？
12. Is the zoo far from here? 动物园离这儿远吗？
13. Will it take long to get to the airport? 去机场要很长时间吗？
14. Go along the street until you come to the traffic lights. 沿这条路一直走到红绿灯那儿。
15. Turn right/left at the second crossing. (Take the second turning on the right/left.) 在第二个十字路口向右/左转弯。
16. Take a No. 46 bus, and get off at the Square. 坐46路公共汽车，在广场下车。
17. It's at the corner of Huaihai Street and Xizang Road. 在淮海街和西藏路的路口。

Exercise 6

一位客人正在餐馆点餐，服务员在帮助他。你将听到以下对话，请按要求填充表格：

Items	Information
What kind of beef	
Vegetable ordered	
Drink ordered	
Things on the table	
Dessert ordered	

Useful Expressions 常用句型

1. Order a meal 点菜
2. Are you ready to order yet, sir? 先生，可以点菜了吗？
3. May I take your order? 您点点儿什么？

Module 1 Basic Practice

4. What would you like? 你想吃点什么？
5. Anything else? 还要别的吗？
6. Would you like some coffee? 来杯咖啡怎么样？
7. May I see your menu, please? 能给我看看菜单吗？
8. I'd like to see a menu, please. 请给我菜单。
9. May I see the wine list, please? 请给我看一下酒单，好吗？
10. What do you recommend? 有什么菜可以推荐的吗？
11. What is your suggestion? 你有没有什么好介绍？
12. Do you have any local specialties? 您这儿有什么地方风味吗？
13. I'd like a cup of coffee, please. 请给我一杯咖啡。

2. Listen to the following telephone dialogues and fill in the blanks.（听电话对话，并填空）

Exercise 1

你是电影院的工作人员，你接到一个咨询电话。

W: Good morning, Cambridge Theater.

M: Good morning. _____?

W: Certainly, sir.

M: _____?

W: At 8:00 p.m. How many tickets would you like?

M: _____.

W: What's your name?

M: _____.

W: Could I have your phone number, please, sir?

M: _____.

W: 7867254.

M: _____?

W: 64 dollars all together, sir.

M: Good.

W: We'll hold the tickets at the door until 7:30.

M: Thank you very much.

Useful Expressions 常用句型

订票基本句型：

1. I'd like to book four tickets, please. 我订四张票。
2. Would you like one way or round trip? 你要单程还是往返票？
3. Round-trip. We'll return on... 往返票。我……返回。
4. Four tickets on... to... and returning to... on... 四张……（日）到……（地点），……

（日）返回到……（地点）的票

订票询问价格：

1. Could you tell me how much it costs to...? 你能告诉我……多少钱吗？

2. The price of a ticket from... to... is... 从……到……的票价是……

Exercise 2

你是酒店的前台服务员，你接到一位客房住客的电话。

M: Good evening. _____?

W: Yes, what can I do for you, sir?

M: Could you please _____?

W: Sure. And your room number, please?

M: _____.

W: All right. An early call at 5：40, Room 1525, Jack Smith.

M: Please don't forget; otherwise, _____.

W: I won't. Have a nice sleep.

M: Thank you. Good night.

Useful Expressions 常用句型

1. Welcome to our hotel. 欢迎到我们酒店来。

2. I hope you will enjoy your stay in our hotel. 希望您在我们酒店过得愉快。

3. Housekeeping. May I come in? 我是客房服务员，可以进来吗？

4. Can you tell me your room number? 您能告诉我您的房间号码吗？

5. May I see your room card? 能看一下您的房卡吗？

6. I'll be with you as soon as possible. 我尽快来为您服务。

7. When would you like me to clean your room, sir? 您要我什么时间来给您打扫房间呢，先生？

8. It is growing dark. Would you like me to draw the curtains for you? 天黑了，要不要我拉上窗帘？

9. We will come and clean the room immediately. 我们马上就来打扫您的房间。

10. I'm coming to change the sheets and pillowcases. 我来替换床单和枕套。

Exercise 3

你是 ABC 公司销售部的前台接待员，你接到远东公司盖茨先生的来电。

W: Good morning, ABC Company. Can I help you?

M: I'm _____. I'd like to _____

　　 I want to _____.

W: Let me check Mr. Chen's diary. Well, would 9：30 Tuesday morning be convenient for you?

M: Yes. _____.

Module 1 Basic Practice

W: Then, next Tuesday morning, at 9:30, Mr. Chen will meet you in the office. And, Mr. Gates, would you mind leaving your contact number?

M: _____.

W: Thank you, sir.

Useful Expressions 常用句型

1. I'm Tom Gates from Far-East Corporation. 我是远东公司的汤姆·盖茨。
2. I would like to make an appointment with Mr. Chen this week. 我想和陈先生约这个星期见面。
3. I want to see him about some details of the contract. 我想和他谈有关合同的细节问题。
4. Let me check Mr. Chen's diary. 让我查查陈先生的日程安排。
5. Would 9:30 Tuesday morning be convenient for you? 星期二上午9点30分对你是否合适？
6. Would you mind leaving your contact number? 你是否介意留下联系电话？

Exercise 4

你是 Mr. Walker 的秘书，你接到 ABC 公司的销售代表的来电。

W: Good morning, Mr. Walker's office.

M: Good morning. This is _____. I have _____.

W: Yes, that's right, Mr. Smith.

M: I'm afraid _____. I have to _____. I will _____. Can we make it _____?

W: Well, would 9:00 next Tuesday morning be convenient for you?

M: Yes, _____.

W: Then next Tuesday morning, at 9:00, Mr. Walker will meet you in his office.

M: Thanks a lot. Bye.

W: Bye.

Useful Expressions 常用句型

1. I have made an appointment with Mr. Walker for 10:00 tomorrow morning. 我已经和沃克先生约好明天上午10点见面。
2. I have to attend an important meeting in New York tomorrow. 我明天必须参加在纽约召开的一个重要的会议。
3. Can we make it some other time? 我们可以另外约一个时间吗？
4. Would 9:00 next Tuesday morning be convenient for you? 下星期二上午9点您是否方便？

英语听打实训教程

Unit 5　Conversation Audio Typing

Aims and Expectations 能力目标

通过本单元的教学和训练，学生应具备以下能力：

1．能听打简单的二人对话。
2．能听打较复杂的三人对话。

在商务活动中，经常需要把大段的口头对话以文字形式记录下来。过去，人们一般使用手写记录信息，但手写记录存在问题较多，如速度慢，字体不规范，以致记录的信息难以辨认等。因此，时下流行的是用电脑记录，并广泛应用于行业或企业举行的重要会议、谈判、律师取证、法庭审讯、记者采访等领域，尤其是在一些不适宜录音的重要场合，这已成为快速、有效的文献和信息保存的主要方式。

本单元由浅入深，模拟实际情景，训练学生听打对话的能力。

Practice Exercise 实训练习

Task

1. Transcribe the three conversations between two people from the recording at 60 words per minute.（听录音，打出三段两人对话，语速为 60 字/分钟）

Exercise 1
Clues：
pick up 接载（某人）

A：_____
B：_____
A：_____
B：_____
A：_____
B：_____
A：_____
B：_____
A：_____
B：_____
A：_____
B：_____

Module 1 Basic Practice

Exercise 2
Clues:

be in touch 保持联系   arrange 安排   accident 意外

A: _____
B: _____
A: _____
B: _____
A: _____
B: _____
A: _____
B: _____
A: _____
B: _____
A: _____

Exercise 3
Clues:

spicy 辛辣的   Beijing Roast Duck 北京烤鸭

A: _____
B: _____
A: _____
B: _____
A: _____
B: _____
A: _____
B: _____
A: _____
B: _____
A: _____
B: _____

2. Transcribe the three conversations among three people recorded at 60 words per minute. （听录音，打出三段三人对话，语速为 60 字/分钟）

Exercise 1
Clues:

envelope 信封   unmarried 未婚的   symbolize 象征着
happiness 幸福   custom 风俗   You bet! 你猜对了!

Wang: _____
David: _____

Wang：_____
Paul：_____
Wang：_____
Paul：_____
Wang：_____
David：_____
Wang：_____
David：_____
Wang：_____
David：_____
Wang：_____
Paul：_____
Wang：_____
Paul & David：_____

Exercise 2
Clues：

replace 代替   resign 辞职   course 课程
Singapore 新加坡   come through 圆满结束   celebration party 庆功宴

Staff A/B：_____
Manager：_____
Staff A：_____
Manager：_____
Staff A：_____
Manager：_____
Staff A：_____
Manager：_____
Staff A：_____
Manager：_____
Staff B：_____
Manager：_____
Staff B：_____
Manager：_____
Staff B：_____
Manager：_____

Exercise 3
Clues：

decent 相当好的   slack 萧条的   financial crisis 金融危机
account 账户   Construction Bank 建设银行

Module 1 Basic Practice

A: _____
B: _____
C: _____
A: _____
C: _____
A: _____
B: _____
A: _____
B: _____
A: _____
B: _____
A: _____
B: _____
A: _____
B: _____
A: _____

3. Transcribe the two conversations among several people recorded at 60 words per minute.（听录音，打出两段多人对话，语速为 60 字/分钟）

Exercise 1

Clues：

go around 到处转转　　　　　　　　staff 职员
show sb. around 带某人四处参观　　first things first 重要的事情先来
Office Manager 办公室主任　　　　Personal Assistant 私人助理

A: _____
B: _____
A: _____
B: _____
C: _____
A: _____
C: _____
B: _____
A: _____
C: _____
B: _____
(In the office)
A: _____
D: _____

英语听打实训教程

B: ___
A: ___
E: ___
F: ___
A: ___
B: ___
F: ___

Exercise 2
Clues:

tentative 暂定的　　　　itinerary 旅行日程　　　considerate 体贴的
agenda 议事日程　　　　expire 到期　　　　　　propose 提议，建议
objection 反对　　　　　cafeteria 自助餐厅　　　compile 整理，编排
Ming Tombs 明十三陵（北京著名景点）

A: ___
B: ___
A: ___
C: ___
A: ___
C: ___
A: ___
C: ___
A: ___
B: ___
A: ___
D: ___
A: ___
D: ___
A: ___
D: ___
A: ___
B: ___
C: ___
D: ___

Test

This part is to test your audio typing skills. You are required to listen and type the following conversation. Assess yourself by the criteria followed.（测试听打能力。听打出一个对话，用后面的评分标准表给自己评分）

Clues:

Sales Department 销售部　　　salad spinner 蔬果沙拉脱水器　　　lettuce 莴苣，生菜

Module 1 Basic Practice

part(s) 组装零件，部件　　　pasta 意大利面
gadget 方便家里或办公室日常工作的小机器
department store 百货公司　　　processor 处理器

A: _____
B: _____
A: _____
B: _____
C: _____
B: _____
C: _____
B: _____
D: _____
B: _____
A: _____
B: _____
D: _____
A: _____
D: _____
B: _____
D: _____
B: _____
A: _____
B: _____

等级	评分标准
优（90~100 分）	1. 指法正确无误； 2. 准确率达到 90% 以上（错误不超过 20 个）。
良（80~89 分）	1. 指法错误不超过 2 个； 2. 准确率达到 85% 以上（错误不超过 33 个）。
中（70~79 分）	1. 指法错误不超过 3 个； 2. 准确率达到 80% 以上（错误不超过 44 个）。
及格（60~69 分）	1. 指法错误不超过 4 个； 2. 准确率达到 75% 以上（错误不超过 50 个）。
不及格（60 分以下）	1. 超过 4 个指法错误； 2. 准确率低于 75%（错误超过 50 个）。

Module 2　Advanced Practice

Unit 6　Memos and Notices

Aims and Expectations 能力目标

通过本单元的教学和训练，学生应具备以下能力：
1. 能使用 Word 文件编辑工具对文字进行编辑。
2. 能听懂录音，并摘录主要信息。
3. 能把摘录的主要信息写成通知。
4. 能把摘录的主要信息写成备忘录。

Word Processing Word 编辑

Word 是一个文字处理软件，主要用于文字的编辑与排版，是 Office 办公软件包中最常用的办公软件之一。它是目前世界上最流行的文字编辑软件。有了它，我们不仅可以编排出精美的文档，而且大大方便了编辑和发送电子邮件，编辑和处理网页等工作。Word 最常用的工具包括：

1. 文字格式

格式工具栏：本栏包括"样式"、"字体"、"字号"、"加粗"、"斜体"、"下划线"、"字符边框"、"底纹"、"字符间距"、"两端对齐"、"居中"和"右对齐"等。

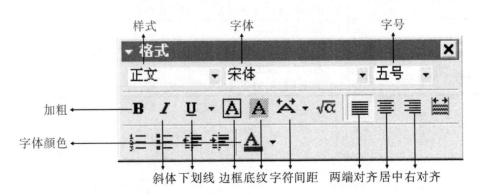

Module 2　Advanced Practice

样式：可以从这里选择各种文本的样式。

字体：正规文本编辑中，中文一般选择"宋体"字体，英文一般选择"Times New Roman"和"Arial"这两种字体，因为这些字体在文本是显示得最清楚的。

字号：字号是指文字的大小。有两种标识方法，一种是中文字号，如一号、二号……号数越大，字越小；另一种是磅值，用数字标识，磅值越大，字越大。

字形：包括"加粗"、"斜体"和"下划线"。

对齐方式：文本的对齐方式有"左对齐"、"右对齐"、"两端对齐"和"居中"。

字体颜色：通过此选项可随意改变文本中字体的颜色。

字符边框和底纹：简单的边框和底纹可使用工具栏中的按钮来编辑，复杂的边框和不同颜色的底纹可通过单击"格式"菜单中的"边框和底纹"，在弹出的对话框中选择"边框"标签，设置边框的类型、线型、颜色等；选择"底纹"标签，即可选择不同颜色的底纹，完成后单击"确定"按钮。见下图：

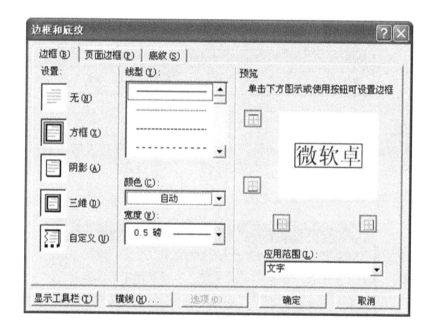

假如要对字体进行更多复杂的编辑，可以通过打开字体对话框来实现。

打开方式：单击"格式"菜单下的"字体"。字体对话框有"字体"、"字符间距"、"文字效果"三个标签。见下图：

英语听打实训教程

2. 段落格式

要对段落的格式进行进一步的编辑，可以单击"格式"菜单下的"段落"，打开"段落对话框"来进行。段落对话框有"缩进和间距"、"换行和分页"、"中文版式"三个标签。见下图：

Module 2　Advanced Practice

（1）设置段落缩进

　　缩进可以通过拖动横向标尺上的缩进按钮进行设置，不过这种方法的缩进量不准确，易造成文字错位，所以缩进最好采用"段落"对话框进行设置。

　　方法：把光标放置在要设置缩进的段落中，或者选择各个需缩进的段落，单击"格式"菜单下的"段落"，选择"缩进和间距"标签，选择需要的缩进方式，并设置缩进量，单击"确定"按钮。

　　缩进方式：①左缩进和右缩进：只需设置缩进量；②段首缩进和悬挂缩进：在"特殊格式"下拉式列表框中进行选择，在旁边的"度量值"框中设置缩进量。

（2）设置段落间距

　　把光标放置在要设置段落间距的段落中，单击"格式"菜单下的"段落"，选择"缩进和间距"标签，在"段前"和"段后"框中设置与上一段和下一段之间的距离，设置完成后单击"确定"按钮。

　　提示：如果同时设置多个段落的间距，可选择这些段落，设置时一般只设置"段前"或"段后"距离中的一个即可。

（3）设置行距

　　把光标放置在要设置行间距的段落中，单击"格式"菜单下的"段落"，选择"缩进和间距"标签，在"行距"的下拉式列表框中选择需要的行距大小，在旁边的"设置值"框中设置行距的磅值，设置完成后单击"确定"按钮。

3．项目符号和编号

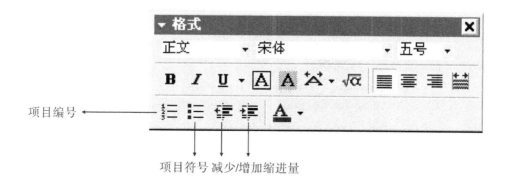

给文本加入项目符号和编号可以通过工具栏中的按钮来编辑。而要编辑多级符号，可使用工具栏中的"减少缩进量"和"增加缩进量"两个按钮来进行。

如果要加入复杂的项目符号和编号，可以通过打开其对话框来进行编辑。

打开方式：单击"格式"，选择"项目符号和编号"。"项目符号和编号"对话框有"项目符号"、"编号"、"多级符号"和"列表样式"四个标签。见下图：

4．表格编辑

（1）插入表格

单击"表格"，选择"插入"，即会弹出以下对话框：

（2）绘制表格

单击"表格"，选择"绘制表格"，即会弹出以下对话框：

Module 2　Advanced Practice

5．页眉和页脚的设计

要插入页眉和页脚，可单击"视图"，选择"页眉和页脚"，即会弹出以下对话框。只要在相应的位置输入页眉的内容，然后单击关闭，即完成页眉的输入。

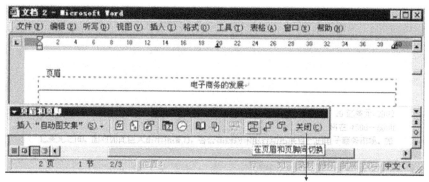

输入完成后按"关闭"

如果同时还要加上页脚，可以按页眉页脚的切换键，见下图：

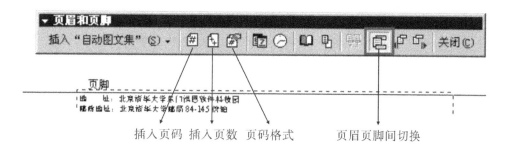

插入页码还可以通过单击"插入"，选择"页码"，使之弹出"页码"对话框来完成。见下图：

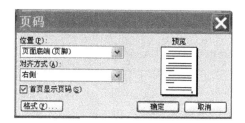

Task

Type the following passage, and compile it as required.（请打出下列段落，并按后面的格式编辑要求对其进行编辑）

Plan for Harrison Electronics

This is the <u>rough plan</u> for Harrison Electronics. The board of directors made a final decision that the company plans to open a new plant in China the next year. The purpose is to develop the full-color videophone. The estimated total investment will reach *$85* million within three years. Based on the report from our R&D division, the total sales will be *$35* million each year once the new plant starts production. We are going to have 1,500 employees in the new plant. According to the work schedule, the first step is to visit China and hold negotiations with the Chinese government. The next is to collect data and the third step is for our researchers to analyze this data and put together a draft proposal here in the headquarters. The final report will be presented to the management in January.

A Proposal for the New Plant
Objective: To open a new plant to develop the full-color videophone in China.
The total estimated investment: *$85* million
The total estimated sales: *$35* million
The number of employees: 1,500
The schedule: 1. To visit China and hold negotiations with the Chinese government. 2. To collect data. 3. Our researchers will make data analysis and write the draft here at the headquarters. 4. The final report will be presented to the management in January.

格式编辑要求：
1. 字体：使用"Times New Roman"字体。
2. 字号：题目字号为四号，正文字号为小四。
3. 字形：按照原文的格式，分别对相关文字进行加粗、斜体和下划线的编辑。
4. 字体颜色：题目字体颜色为绿色，正文为黑色。
5. 行距：1.5倍行距。
6. 正文：首行缩进。
7. 编号：使用"项目符号和编号"对话框，加入绿色字体的阿拉伯数字编号，并使用"增加缩进量"按钮使之往内缩进。
8. 页眉和页脚：用文章的标题作为页眉，页脚处键入页码。

Basic Structure of Notice 通知的写作

1. 英文通知的基本写作技巧

通知（Notice）是上级对下级、组织对成员布置工作、传达情况或告诉公众某种事情

Module 2　Advanced Practice

等时使用的一种应用文体。

　　通知格式的写法有点类似于书信的写法，其格式为：

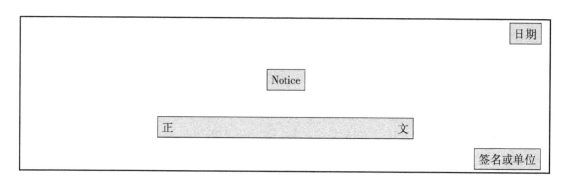

　　通知是传达将要做的事情，因此，写通知多用一般现在时和将来时态。书写通知的正文时，语言应简洁明了，把通知的对象、事由、时间、地点、内容有条理地说清楚即可。简单地说，就是"某人（单位）某时在某地干某事，加上注意事项"即可。当然，必须注意句子之间的安排，使之符合逻辑，条理清楚。

　2. 常用句型

英文通知通常会涉及以下句型：

（1）May I have your attention, please? 请注意。

（2）I have something important to tell you. 我有重要的事情要告诉你们。

（3）I have got an announcement to make. 我要宣布一件事情。

（4）I am told that a speech contest will be held at the English Competition this summer vacation. 我听说这个暑假要举办一场英文演讲比赛。

（5）Everyone is welcome to attend the party. 欢迎大家参加这个晚会。

（6）Please bring along your pens and notebooks. 请带着笔和笔记本。

（7）All the students are requested to attend the meeting to be held in the meeting-room on the third floor at 6：30 p.m. this Wednesday. 要求所有的学生都参加这个周三下午6点半在三楼会议室举行的这个会议。

（8）Don't miss the chance. Please get there on time. 不要错过这个机会，请准时到场。

（9）Please get everything ready ahead of time. Don't be late. 请提前做好准备，不要迟到。

（10）There will be a lecture on pollution given by Professor Wang from Beijing University in the school hall at 8：00 tomorrow morning. 明天上午8：00在学校会堂由北京大学的王教授做一个关于污染的报告。（后可加上：希望同学们准时参加。请带上笔和笔记本。即：I hope all the students can attend the lecture on time. Please take your pens and notebooks.）

（11）The lecture will last 3 hours. 讲座将持续3个小时。

（12）The lecture will begin at 3：30 p.m. and end at 5：00 p.m. 这个讲座将在下午3：30开始并在下午5：00结束。

（13）After the report, we'll have a discussion about the topic. 报告之后我们将对这个话题进行讨论。

Part III

Practice Exercise 实训练习

Task

Listen to the recordings, take down the key information and complete the following four notices.（听录音，摘录其主要信息，并撰写四份通知）

Exercise 1
Clues：

monitor 班长　　　　　　meeting hall 会议大厅　　　　Christmas Eve 平安夜
performance 表演　　　　English play 英文戏剧　　　　poem 诗歌

Dec. 21st, 2009

Notice

An English Christmas evening party is to be held _____
_____
_____
Please get everything ready and be at the party on time.

Lily (Monitor)

Exercise 2
Clues：

the sports meet 运动会　　put off 推迟　　　　　　require 要求
as usual 如常　　　　　　permit 允许　　　　　　　basketball court 篮球场
match 比赛　　　　　　　fantastic 极好的

Apr. 10th, 2009

Notice

The sports meet _____
_____
_____
There will be a basketball match between No. 1 Middle School and our school at 5：00.

Alex (Secretary of the Sports Union)

Module 2　Advanced Practice

Exercise 3
Clues：

secretary 秘书　　　　　　Head Office 总部　　　　　announce 宣布
colleague 同事　　　　　　Sales Manager 销售经理　　expert 专家
marketing 市场营销

```
                                                            March 8th, 2010
                              Notice
  A new Sales Manager (Sally Smith) _____
  _____
  _____
  Thank you for your cooperation.
                                        Joyce White (Secretary of the Head Office)
```

Exercise 4
Clues：

export 出口　　　　　　　　order 订单　　　　　　　　raise 改善，提高
training course 培训课程　　deal with 处理

```
                                                          August 18th, 2009
                              Notice
  Recently we received a large number of export orders. _____
  _____
  _____
  Thanks for your cooperation.
                                                    Ken Wang (Training Manager)
```

Part Ⅳ

Basic Structure of Memos 备忘录的结构

备忘录是一种录以备忘的简易公文，它主要用于公司内部各部门之间的书面联系，传递信息，提醒或督促对方，或表明自己的看法。当然，备忘录也可以应用于公司与公司之间，甚至国家与国家之间。备忘录通常由抬头、发函人姓名、收函人姓名、日期、事由和正文等组成。另外，也可以根据需要增加签名、附件、副本抄送等部分。

英语听打实训教程

1. 抬头：大多数公司都有印好抬头的备忘录专用纸。如用白纸，只需在信纸的正上方注明 MEMO/Memorandum 即可。
2. 日期：现成的备忘录纸上往往印有"Date"，只需填上日期、月份即可，名称尽量用全拼或缩写，月份避免用阿拉伯数字，以免混淆。
3. 收函人姓名：现成的备忘录上常印有"To"字样，如没有就自己写上。然后填上收函人姓名，姓名前应加以礼貌性称呼，有时也可以在姓名后加上其职务及部门。如收函人较多，可以在收函人栏注明"see distribution"，然后在正文后列出收函人姓名。
4. 发函人姓名：写在"From"后面，一般不必带礼貌性称呼，但可以加注职称或职务。
5. 标题：也称为事由，写在"Re"或"Subject"字样后面，用一个短语甚至一个单词对备忘录中的事情做一简要概括，使收函人能够一目了然，及时处理。
6. 正文：根据内容，言简意赅。
7. 签名：由于上边已有发函人姓名，结尾时可以没有签名。但如事关重要，可签名。如果该备忘录是代替上司发出的，应请求上司签名。
8. 附件：备忘录如附有其他材料，应在正文之后加注说明，写在"enclosed"字样后面。
9. 副本抄送：如该备忘录须抄送收函人以外的人员或部门，应在备忘录正文后加以说明，写在"Cc"字样后面。

Practice Exercise 实训练习

Task

Listen to the recordings, take down the key information and complete two memos. （听录音，摘录其主要信息，并撰写两封备忘录）

Exercise 1

Clues：

Sales Accountant 销售会计师　　complaint 投诉　　　　Singapore 新加坡

model 样板　　　　　　　　　　issue 事情　　　　　　apologize 道歉

Module 2 Advanced Practice

Memo

To: _____
From: Mr. White (_____)
Date: April 15th, 2010
Subject: Complaint from a customer in Singapore
Cc: Mr. King (the manager)

_____

_____

I hope this suggestion is satisfactory.

Exercise 2

Clues:

salesman 销售员   be satisfied with 对……感到满意   offer 提供
quit the position 辞职   salary 薪水

Memo

To: _____
From: _____
Date: May 6th, 2010
Subject: _____

_____

_____

_____ I therefore write this memo so as to resign from my job. I am giving the required two-week notice.

Unit 7　Business Letters

Aims and Expectations 能力目标

通过本单元的教学和训练，学生应具备以下能力：
1. 能通过听录音摘录主要信息。
2. 能运用听下来的信息写出相应的书信。

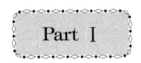

Basic Structure of English Business Letters 英语商务信函的基本结构

1. 英语商务信函的必要组成部分

标准的英文商务信函应该由信头、信文和信尾三个部分组成。

（1）信头，包括发信人地址、发信日期和收信人姓名和地址、称号、至 xxx 栏及事由。很多公司使用公用信纸，上面已印有公司名称、通信地址、邮政编码、电话号码、传真号码和电子邮箱等内容。

（2）信文，也称为正文，为信函的主要内容。正文是商务信函的主体或核心，它向收信人传达有关商业信息，以便达成商业交易或解决商业纠纷。在写作商业信函时，要重视遣词造句，避免语法错误；重视传达信息的准确性，避免含糊疏漏；注重礼貌友好，避免命令威胁。

（3）信尾则包括结尾敬辞和信末签名。结尾敬辞应该写在正文的后面，另起一行，但在签名的上方。签名即发信人的姓名。在西方，签名是有法律效用的，它代表一个人的信用。名字应该由发信人用钢笔亲自签署，不必考虑信件是手写的还是用电脑打印的，也不用顾虑收信人是否能认出你的名字，因为在手签的下一行，还要用正楷（或电脑）写清楚你的名字、职务及公司名称。

2. 其他组成部分

根据内容的不同，商务信函还有一些特殊组成，可根据需要取舍。

（1）主题行：常标记为 Re：或 Subject，用简单的短语表达信文的主要内容，位于称呼之下。

（2）附寄标志：如随信寄有其他物品，应在信尾签名下面标注。其标记通常为 encl. 或 enclosed，并在其后标明附寄物品的名称和数量。

（3）分送标志：当发信人希望收信人得知该信发至除他以外的其他人时，应加注分送标志，其标志为 Cc：，并在后面注明其他收信人的名称。

（4）附言：信件完成后又临时想起的不太重要的事情，可用附言提及。其标志为 PS,

Module 2　Advanced Practice

即 PostScript 的缩写。但注意不要本末倒置，把重要的事情写到附言中去。

见下面信件样式：

Actex Co. Ltd. 发信人名址

1-2-3 Hamayama-dori
Hyogo-ku
Kobe City, Hyogo Pref.
Japan

June 3rd, 2009 写信日期

Mr. Teo Pin 收信人名址
Singapore Moulds and Tools Centre Ltd.
Blk 6020
Ang Mo Kio Industrial Park 3
Singapore 569574

Dear Mr. Teo, 称呼

Re：Arrangements regarding Mr. Tang's visit 事由

I have received your letter concerning your wish to send Mr. Tang to visit us at Actex Co. Ltd. I completely agree that a visit by one of your senior staff would be beneficial to both of our organizations.

Please let me know if this arrangement is satisfactory. 正文

Yours sincerely, 结尾敬辞

Ken Fukuzawa 手写签名

Ken Fukuzawa 打印发信人姓名

Public Relations Officer 职务

Cc：Mr. Kukubo (General Manager) 抄送
Mrs. Smithson (Marketing Manager)

Practice Exercise 实训练习

Task

1. Listen to the recordings, take down the key information and complete three letters of invitation.（听录音，摘录其主要信息，并撰写三封邀请信）

Useful Expressions 常用句型

1. I'm delighted that you have accepted our invitation. 我很高兴您能接受我们的邀请。

47

2. I look forward to hearing from/seeing you. 我期待得到您的回复/我期待见到您。
3. Look forward to your favorable reply. 期待您的佳音。
4. We would very much like to have you give a lesson to our members. 我们非常希望请您来给我们的职员上一次课。
5. We would like to invite you to an exclusive presentation of our new products. 我们希望邀请您来参观我们的一场新产品的专场展示。

Exercise 1

Here is a letter with some of the information missing, please fill in the blanks while listening to the tape.（下面是一封不完整的书信，请听取信息把它补充完整）

Clues：

celebrate 庆祝　　　　　　　graduation 毕业

Dear Anna,

_____
_____
_____
_____ .

Look forward to your favorable reply.

Yours faithfully,
Mary

Exercise 2

Clues：

chairman 主席　　　　　　　Student Council 学生会
Guangming Financial Vocational School 光明金融职业学校
professor 教授　　　　　　　speech 演讲　　　　　Western cultures 西方文化

Dear Professor Smith,

I am the chairman of the Student Council in Guangming Financial Vocational School. _____
_____
_____ .

Look forward to your favorable reply.

Yours faithfully,
Wilson

Module 2 Advanced Practice

Exercise 3

Clues:

dance party 舞会   neighbourhood 社区   Summer Club 夏日会所

```
Dear Ellen,
_____
_____
_____
_____.

I would love for you to attend, so please let me know your decision.

Yours faithfully,
Helen Liu
```

2. Listen to the recordings, take down the key information and complete three letters of thanks.（听录音，摘录其主要信息，并撰写三封感谢信）

Useful Expressions 常用句型

1. I am writing this letter to thank you for your warm hospitality to me. 我写这封信是表达对您热情招待的感谢。
2. I am writing to show my thanks to you. 我写这封信来表达我对您的谢意。
3. I would like to take this opportunity to express my heartfelt gratitude to you. 我想借此机会向您表示我衷心的谢意。
4. I am writing to tell you how very much I enjoyed the days at Chicago. 我想借此信告诉您我在芝加哥度过了一段很开心的日子。
5. With thanks again and best wishes to you. 再次向您致以谢意和祝福。

Exercise 1

Here is a letter with some of the information missing, please fill in the blanks while listening to the tape.（下面是一封不完整的书信，请听取信息把它补充完整）

Clues:

trip 旅途   hometown 家乡   show sb. around 带某人四处参观
unforgettable 难忘的   friendship 友谊

英语听打实训教程

Dear Jane,

_____
_____
_____
_____ .

With kind personal regards,

Faithfully yours,
Li Ping

Exercise 2

Dear Mr. Brown,

_____
_____
_____
_____ .

Sincerely yours,
Helen Wang

Exercise 3

Clues:

first prize 一等奖 English Competition 英语竞赛 win the first place 赢得一等奖
rapid 快速的 achievement 成就,成绩 owe...to sb. 把……归功于某人

Dear Mr. Smith,

_____
_____
_____
_____ .

Best wishes to you.

Yours faithfully,
Wang Ling

Module 2 Advanced Practice

3. Listen to the recordings, take down the key information and complete two letters of establishing business relationship. （听录音，摘录其主要信息，并完成两封建立贸易关系的信）

Useful Expressions 常用句型

1. Having obtained your name and address from... 我们从……得知贵方的名称和地址
2. On the recommendation of... 承蒙……的介绍
3. We owe your name and address to... 我们从……处得知贵方的名称和地址。
4. We are writing to you in the hope of establishing business relations with you. 现写信给您想与贵公司建立贸易关系。
5. From alibaba.com, we have learned that you are a... in France, and at present you are in the market for... 通过阿里巴巴网，我们得知贵公司是法国主要的……，目前正想购买……。
6. We are an enterprise specializing in... 我们是一家专营……的厂商。
7. Our products enjoy great popularity in the world market. 我们的产品在国际市场享有盛誉。
8. Enclosed is a copy of our new products' catalogue. 随函附寄一份我们新产品的目录。
9. We can offer a wide range of... with sensible prices and high quality. 我们能以具有竞争力的价格和出色的品质提供广泛多样的……。
10. We are well acquainted with the market conditions in Southeast Asia. 我们对东南亚的市场情况非常熟悉。

Exercise 1

Here is a letter with some of the information missing, please fill in the blanks while listening to the tape. （下面是一封不完整的书信，请听取信息把它补充完整）

Clues:

alibaba.com 阿里巴巴网站　　　　importer 进口商　　textile 纺织品
specialize in 专门从事，专攻　　　establish business relations 建立贸易关系
catalogue 产品目录　　　　　　　popularity 受欢迎

Dear Sirs,

_____
_____
_____
_____
_____.

Please let us have your specific enquiry if you are interested in any of the items listed in the catalogue. We shall make offers promptly.

Look forward to your favorable reply.

Yours faithfully,
Huang Wei
Encl. As Stated

Exercise 2
Clues：

importer 进口商 Chinese Chamber of Commerce 中华商会
leather products 皮革制品 quotation 报价

Dear Sirs,

_____
_____.
_____
_____.

We are looking forward to receiving your early reply.

Yours faithfully,
Joanna Wood

---

4. Listen to the recordings, take down the key information and complete two sales letters.（听录音，摘录其主要信息，并撰写两封销售信）

Useful Expressions 常用句型

1. ... has been highly recognized by... for its much lower price and good quality. ……（产品）因其价格低廉和良好的质量得到了……的认可。
2. In order to popularize these products, all the prices... 为了推广这些产品，所有的价格……。
3. For the detailed information, please find the enclosed catalogue. 详情见随函附寄的目录。
4. It is universally acknowledged that the quality of our goods is of first-grade. 大家一致公认我方产品的质量是一流的。
5. We are a professional exporter of toys in China with more than 10 years' experience and have already set up a good reputation in the world market, due to our good management system and excellent after-sale service. 我方具有十多年的经验，是中国专业的玩具出口商。由于良好的管理制度和出色的售后服务，我们已在国际市场建立了良好的声誉。
6. Our products are all supplied by the first-class manufacturers of this country, and so we are in a good position to serve your customers with the most reliable quality of the line you suggest. 我方商品均由本国一流厂商提供，因此，我方有条件向贵方的客户提供质量最可靠的商品。

Exercise 1
Here is a letter with some of the information missing, please fill in the blanks while listening to the tape.（下面是一封不完整的书信，请听取信息把它补充完整）

Clues：

specialize in 专门从事于 bicycle 自行车 recommended 得到承认的

Module 2　Advanced Practice

customer 客户　　　　quality 质量　　　　discount 折扣
catalogue prices 目录价，标价

Dear Sirs,

_____
_____
_____
_____
_____.

I believe there will be a good chance of starting business and even establishing partnership between us. And if you have any question or want more information, please feel free to contact me.

Yours faithfully,
Wang Ming
Sales Manager

Exercise 2

Clues：

sales representative 销售代表　　Import & Export Co. 进出口公司　　demand 需求
Christmas doll 圣诞玩偶　　　　stuff 填充　　　　　　　　　　　　polyester 涤纶
international safe standard 国际安全标准　　　　　　　　　　　　　client 客户
requirement 要求

Dear Sirs,

_____
_____
_____
_____
_____.

For the detailed information, please find the enclosed catalogue No. 123.

Your immediate reply would be highly appreciated.

Yours faithfully,
Nanjing Green Leaves Import & Export Co., Ltd.
Mary Li

Encl. As Stated

5. Listen to the following two dialogues, take down the key information and complete two letters of apology. （听以下两段对话，摘录其主要信息，并撰写两封道歉信）

Useful Expressions 常用句型

1. We regret to tell you that, owing to..., we are unable to... 我们很抱歉告诉你，由于……，我们无法……。
2. As a result of..., we couldn't... for which we express our deep regret to you. 由于……，我们无法……，在此我们表示深深的歉意。
3. We regret to say that we are out of stock of such articles as you described. 我方对于你方所述商品没有存货，对此十分抱歉。
4. It is regrettable to see an order dropped owing to no agreement on prices. 很遗憾因价格未谈妥不能成交。
5. I would like to express my apologies for... 对于……，我深表歉意。

Exercise 1

Clues：

pet toy 动物玩具　　　　out of stock 无现货（库存）　　new stocks 新货
regret 遗憾　　　　　　 owing to 由于　　　　　　　　shortage 缺少
delivery 发货，出货

Dear Sirs,

Thank you for your order of September 10th for pet toys. _____
_____
_____
_____.

Thanks for your understanding.

Yours faithfully,
Helen Liu

Exercise 2

Clues：

shipment 货运，出货　　　manufacturer 生产商　　　　fix the date 定下日子
get in touch 联系　　　　 prompt 马上的，立刻的　　　delay 延迟
shipping date 出货期　　　cooperate 合作

Module 2　Advanced Practice

```
Dear Sirs,
Thank you for your e-mail of June 20th. ＿＿＿＿＿＿＿＿＿＿＿＿＿＿＿＿＿＿＿
＿＿＿＿＿＿＿＿＿＿＿＿＿＿＿＿＿＿＿＿＿＿＿＿＿＿＿＿＿＿＿＿＿＿＿＿＿＿＿
＿＿＿＿＿＿＿＿＿＿＿＿＿＿＿＿＿＿＿＿＿＿＿＿＿＿＿＿＿＿＿＿＿＿＿＿＿＿．

As soon as the shipment is completed, we shall call you the shipping advice without delay.

Yours faithfully,
Wang Gang
```

6. Listen to the following two dialogues. Take down the key information and complete two letters of congratulations. （听以下两段对话，摘录其主要信息，并撰写两封祝贺信）

Useful Expressions 常用句型

1. It is a pleasure to congratulate you on. . . . 非常高兴祝贺你在……
2. Again, congratulations and best wishes for continued success. 再次祝贺并祝福你取得更多的成功。
3. I hasten to offer you my hearty congratulations. 谨致上我衷心的祝福。
4. Many congratulations once again on . . . 再次恭喜您……
5. Please accept my warmest congratulations on. . . . 请接受我对您的……诚挚的祝贺。
6. I wish you every success in your future career. 我祝愿您在未来的工作中取得成功。

Exercise 1

Clues：

promotion 升职　　　Deputy Managing Director 副总经理

association 交往　　　be qualified for 有能力，有资格做某事　　　post 职位

```
Dear Mr. Smith,
＿＿＿＿＿＿＿＿＿＿＿＿＿＿＿＿＿＿＿＿＿＿＿＿＿＿＿＿＿＿＿＿＿＿＿＿＿＿＿
＿＿＿＿＿＿＿＿＿＿＿＿＿＿＿＿＿＿＿＿＿＿＿＿＿＿＿＿＿＿＿＿＿＿＿＿＿＿＿
＿＿＿＿＿＿＿＿＿＿＿＿＿＿＿＿＿＿＿＿＿＿＿＿＿＿＿＿＿＿＿＿＿＿＿＿＿＿．
Again, congratulations and best wishes for continued success.

Sincerely yours,
Blake Lee
```

Exercise 2

Clues：

Gold Star Prize 金星奖　　　capable 有能力的　　　CEO 首席行政长官，总裁

英语听打实训教程

appreciate 欣赏
award 奖励
recognition 认可

devote 奉献于
management 管理
noble 卓越的

achievement 成绩，成就
satisfaction 满足

```
Dear Dr. Brown,
_____
_____
_____
_____.

Best wishes

Yours sincerely,
Li Yan
```

7. Listen to the following two dialogues. Take down the key information and complete two letters of urging shipment. （听以下两段对话，摘录其主要信息，并撰写两封催运信）

Useful Expressions 常用句型

1. As... is rapidly falling due, it is imperative that you... without any further delay. 由于……将要到期，你方必须立即……。

2. The above order is urgently needed. Therefore, we must insist on express shipment. 由于急需上述订单所订货物，因此我方要求务必特快装运。

3. To ensure fastest delivery, you are requested to forward the above order by air. 为保证尽快交货，要求空运上述订单货物。

4. We must insist on delivery within the time dated, and reserve the right to reject the goods, should they be delivered later. 我方坚持必须在规定时间交货，如果迟交货我们有权拒收货物。

5. It is imperative that you ship the goods immediately. 你方必须立即装运该货。

Exercise 1
Clues：

come due 到期
contract number 合同号

urge 催促
plastic toy 塑料玩具

anxiety 渴望
state 陈述，规定

Module 2 Advanced Practice

Dear Sirs,

_____
_____
_____
_____
_____.

We are looking forward to your early reply.

Yours faithfully,
Mary White

Exercise 2
Clues:

end user 最终用户   in urgent need of 急需   vessel 货船
Vancouver 温哥华   pick up 收，接   proposal 提议

Dear Sirs,
Referring to the contract No. 456, the date of shipment is in October. _____
_____
_____
_____
_____
_____.

Awaiting your early reply.

Yours faithfully,
Tom Smith

Unit 8　Meeting Minutes

Aims and Expectations 能力目标

通过本单元的教学和训练，学生应具备以下能力：
1. 能听懂录音材料，并摘录主要信息。
2. 能根据摘录信息写成会议纪要。

Part Ⅰ

Meeting Minutes 会议纪要

1. 会议纪要主要内容
（1）name of the organization 组织名称
（2）place, date and time of the meeting 会议时间、地点
（3）whether the meeting is regular or called for a special purpose 是否是例会
（4）name of the person presiding 主持人姓名
（5）a record of attendance 出席人员
（6）a reference to the minutes of the previous meeting 早先会议资料
（7）an account of all reports, motions, or resolutions made 报告、提议或做出决定的理由
（8）date, time and place of the next meeting 下次会议时间、地点
（9）time of adjournment 会议结束时间

2. 会议纪要注意事项
（1）The body of the minutes should contain a separate paragraph for each topic. 每个会议议题分段记录。
（2）The minutes should be signed by the person preparing them. 要有筹备会议者签名。
（3）The minutes should be done in time and brief. Do not use too many abbreviations. 会议纪要要及时，简短，但不能用太多缩略语。

Module 2　Advanced Practice

Practice Exercise 实训练习

Task

1. Listen to the following two dialogues about a teachers' meeting. Take down the key information and complete the two meetings' minutes.（听以下两个教师会议的对话，摘录其主要信息，并撰写两封会议纪要）

Exercise 1

Clues：

call the meeting 召开会议　　celebrate 庆祝　　　　Teacher's Day 教师节
opinion 观点　　　　　　　　experience sharing meeting 经验交流会
headmaster 校长

Meeting Minutes

Time：3：00 p.m., Friday, Sep. 2nd, 2009
Venue：Meeting room
Present：Li Zhaoji　　　Liu Haiyun
　　　　　Wang Ming　　Chen Daming
Mrs. Li held the meeting to discuss the following two matters：
1. _____
2. _____
Suggestion for the first matter：_____
_____
Suggestion for the second matter：_____
Result will be declared：_____

Exercise 2

Clues：

graduating student 毕业班学生　　　　practical training 实训，实习
major 专业　　　　　　　　　　　　put skills into practice 把技能应用到实践中
employee 雇员　　　　　　　　　　　simulated interview 模拟的面试

英语听打实训教程

```
┌─────────────────────────────────────────────────────────────┐
│                    Meeting Minutes                          │
│  Time：3：00 p.m., Tuesday, March 2nd, 2010                 │
│  Venue：Meeting room                                         │
│  Present：Headmaster      Mrs. Gao                          │
│           Mrs. Zhang      Mr. Zhao                          │
│  The meeting is held to discuss：_____     │
│  Suggestions in the meeting：                                │
│  1. _____          │
│  2. _____          │
│  3. _____          │
│  4. _____          │
└─────────────────────────────────────────────────────────────┘
```

2. Listen to the following two dialogues about a students' meeting. Take down the key information and complete two meetings' minutes.（听以下两个学生会议的对话，摘录其主要信息，并撰写两封会议纪要）

Exercise 1

Clues：

dormitory 宿舍   take notes 做笔记   discussion 讨论
Baiyun Mountain 白云山   climb mountain 爬山   fresh air 新鲜空气
Yuexiu Park 越秀公园   Five Goats 五羊   symbol 标志

```
┌─────────────────────────────────────────────────────────────┐
│                    Meeting Minutes                          │
│  Time：11：00 a.m., Monday, March 8th, 2010                 │
│  Venue：Classroom                                            │
│  Present：Xiao Yan      Min Yi        Zhang Xin             │
│           Liang Hai     Li Jun        Hai Di                │
│  Minutes taker：_____              │
│  Item discussed in the meeting：_____      │
│  Opinion from Rooms 305, 306 and 308：_____      │
│  Opinion from Room 307：_____      │
│  Result of the meeting：_____      │
└─────────────────────────────────────────────────────────────┘
```

Exercise 2

Clues：

badminton 羽毛球   match 比赛   complain 抱怨
English Vocabulary Match 英语单词竞赛   learning experience 学习经验

Module 2 Advanced Practice

Meeting Minutes

Time: 5:00 p.m., Thursday, March 4th, 2010
Venue: Classroom
Present: Monitor   Liu Jun
         Wang Xuan   Yang Qiang
         Other two leaders
Topic of the meeting: _____
Result of the meeting (four activities in the new term):
1. _____
2. _____
3. _____
4. _____

3. Listen to the following two dialogues about a company's meeting. Take down the key information and complete the two meetings' minutes. （听以下两个会议的对话，摘录其主要信息，并撰写两封会议纪要）

Exercise 1
Clues:

invitation 邀请   opportunity 机会   give a speech 发表演讲
data 数据

Meeting Minutes

Time: 11:00 a.m., Monday, Feb. 8th, 2010
Venue: Meeting room
Present: _____
Items discussed in the meeting:
_____
Points discussed and decisions made:
1. _____
2. _____
3. _____
4. Next meeting: _____

Exercise 2
Clues:

sales figures 销售额，销售数字   purpose 目的   drop 下降，下滑
competition 竞争   sales people 销售员   motivate 激发，诱发
bonus system 分红制度   quota 定额

英语听打实训教程

Meeting Minutes

Time: 9:00 a.m., Monday, Feb. 8th, 2010

Venue: Meeting room

Present: Jack   Amy

           Mary   Dick

Two purposes of the meeting:

1. _____

2. _____

Two reasons for the drop in sales:

1. _____

2. _____

Result: _____

Module 3   The Project

Judy's Day

Judy is a beautiful girl, 23 years old. She works in Kenny and Johnson's Company as a secretary of the General Manager. Every morning, she is the first to arrive at the office and always gets everything ready before the manager goes to work.

8:30 a.m.

Today, after everything was done as usual, she rested herself comfortably on her chair with a cup of coffee, waiting for the manager to work.

9:00 a.m.

Mr. Smith, the manager arrived at the office.

"Good morning, Mr. Smith." Judy stood up and greeted him with a smile.

"Good morning, Judy." Mr. Smith stopped at her table and said, "Come with me to my office."

At the manager's, Mr. Smith said...

(Now, listen to the conversation between Mr. Smith and Judy.)

Coming back to her place from Mr. Smith's office, Judy turned on her computer and began to write a notice.

Task

Notice

_____
_____
_____
_____
_____
_____

10: 00 a. m.

Mr. Smith received a complaint letter from the company's chief customer—Kremel's Industry, saying that they had received goods with faulty quality. So, Mr. Smith called an urgent meeting to solve the problem. Steven Lee—Sales Manager, Johnny Brown—Production Manager, Simon Edgar—Supervisor of R&D Department, Susan Wang—Merchandiser attended the meeting. Judy was also there to take the minutes.

(Now, listen to the conversation at the meeting.)

The meeting ended at 11: 00 a. m. and Judy began to write the meeting minutes.

Meeting Minutes

12: 00 a. m.

At noon, Judy had her lunch at the restaurant nearby with the other three girls from the office—Joanne, Maggie and Mary. The four lovely girls ate and chatted happily. They stayed there for about an hour and then returned to the office. She resumed working at 2: 00 in the afternoon.

2: 30 p. m.

The telephone on the table rang and Judy answered.

"Hello, Judy's speaking." She said.

"Judy, come to my office for a moment." Mr. Smith's voice came from the other end of the line.

Judy entered Mr. Smith's office and the manager handed her a piece of paper with a hand-written letter in it.

"This letter is to reply to Mr. Kremel's complaints I received this morning. Please type it for

Module 3 The Project

me and send it out before 4 o'clock.", said Mr. Smith.

"No problem, sir." Judy answered.

Judy stepped out of the office with the letter.

3

(Judy – please type & send out this letter before 4pm – Thanks, Smith)

Mr. Edward Kremel
Gailey Asia Ltd.
25/F, Omega Plaza
32 Dundas Street
Mongkok, Kowloon

Dear Mr. Kremel, — leave space here
Subject: ORDER No. 125 – poor quality materials

Thank you for your letter this morning stating that you have received products of poor quality from us. I called an emergency meeting to look into the matter and find it is fully our responsibility due to the fact we ordered and accepted materials from a different supplier.

Please accept my sincere apologies for this inconvenience to you.

Would you please return the goods to our factory; freight at our cost, as soon as possible? We will arrange the production of new products for you immediately. If you disagree, please let us know and we will do our best to find another solution for you.

We thank you for your business and hope we can continue looking after & supplying you with our products in the future.

Yours sincerely,
John Smith
General Manager

英语听打实训教程

3：00 p. m.

Judy finished typing and editing the letter accordingly and took the printed version back to Mr. Smith's office.

"Mr. Smith, here is the printed letter. Would you please check it again and sign your name underneath?", said Judy.

"Sure." Mr. Smith had a quick look at the letter and then signed his name.

"Thank you." Judy took back the letter and went out.

She put the letter into an envelope and sent it out.

4：00 p. m.

The telephone on the table rang again.

(Now, listen to a phone conversation.)

Memo

5：30 p. m.

The bell in the office rang punctually and a busy day was over. Judy packed her things happily and quickly. When she hurried downstairs, she found John, her boy-friend who worked in the company nearby, standing there and waiting for her. She smiled a happy smile and walked over to him. Then, they left happily, hand in hand.

Module 4 English Audio Typing Practice

Task

Listen to the following dialogues, take down the key information and fill in the form.
（听对话，摘录主要信息，并填表）

莉莉在看杰克的家庭相册，并谈论起杰克的家庭。你将听到以下对话，请按要求填充表格：

Items	Information
Job of Jack's grandpa	
Job of Jack's grandma	
Job of Jack's parents	
Jim's age	
Location of Ann's school	

Useful Expressions 常用句型

询问他人家庭情况：

1. What does your father/mother/brother/sister do?
   What is your father/mother/brother/sister? 你爸爸/妈妈/哥哥/弟弟/姐姐/妹妹是从事什么工作的？
2. Do you have brothers or sisters? Have you got brothers or sisters? 你有兄弟姐妹吗？

描述家人情况：

1. He is retired now. 他现在退休了。
2. He used to be a doctor. 他曾经是一个医生。
3. She is a housewife. 她是一个家庭主妇。
4. They are all English teachers. 他们都是英语老师。
5. I have an elder brother and a younger sister. 我有一个哥哥和一个妹妹。
6. I have a brother, but no sisters. 我有一个哥哥但没有姐妹。
7. I am an only child. 我是独生子女。
8. My brother is married. 我哥哥已经结婚了。

9. My sister is single. 我姐姐单身。
10. He is a college student now. 他现在是一个大学生。
11. She studies in a middle school in New York. 她在纽约的一所中学读书。

汤姆和玛丽正在讨论最近的天气。你将听到以下对话，请按要求填充表格：

Items	Information
Weather today	
Weather Mary likes	
What is Tom going to do this weekend?	
Weather in London now	
Season in Sydney now	

Useful Expressions 常用句型

询问天气：

1. What's the weather like in London in this season? 伦敦在这个季节是什么样的天气？
2. How about the weather in Sydney now, Mary? 玛丽，现在悉尼的天气怎么样？
3. What's the weather like today? How is the weather today? 今天的天气怎么样？
4. What does the weather forecast say? 今天的天气预报怎么说的？
5. What's the temperature today? 今天的气温如何？

描述天气：

1. It says in the newspaper that the weather today will be rainy and cold again. 报纸说今天的天气又有雨，而且寒冷。
2. And sometimes it is snowy for a few days. 有时会持续好几天下雪。
3. Pretty hot. Now it is summer there. 非常炎热。那里现在是夏天。
4. It looks as if it is going to rain. 看来要下雨了。
5. It's sunny. / It's clearing up. 天晴了。
6. There will be a storm tomorrow. 明天要有一场风暴。

描述个人对天气的喜恶：

1. Well, to tell you the truth, I don't like this kind of weather. 老实告诉你，我不喜欢这样的天气。
2. I prefer it warm and dry. 我喜欢温暖干燥的天气。
3. Weather like this makes me feel tired. 这样的天气使我感到疲倦。
4. A lovely day, isn't it? 天气真好，不是吗？
5. I hope it stays fine. 我希望天气一直晴朗。
6. It's good to see the sun again. 真高兴又看到太阳了。

Module 4 English Audio Typing Practice

7. It's rather changeable, isn't it? 天气变化无常，不是吗？

比尔和玛丽在路上遇到，玛丽询问比尔的近况。你将听到以下对话，请按要求填充表格：

Items	Information
How is Bill recently?	
Bill's task recently	
Mary's advice	
Club's name	
Why is climbing mountains good sport?	

Useful Expressions 常用句型

1. How are you these days? 你近来好吗？
2. Not so well recently. I always feel tired and sick. 最近感觉不是很好。我经常感到疲劳，不舒服。
3. What's the matter with you? 你怎么啦？
4. I think I must be over-worked. 我想我是劳累过度了。
5. Don't work yourself so hard. 别过分操劳了。
6. I think you'd better have a good rest now. 我想你最好先好好休息一下。
7. I think you should do some exercise in your spare time. 我想你应该在空闲时间里做点运动。
8. I think climbing mountains is an interesting sport, and the fresh air is good for our health. 我觉得爬山是一种很有趣的运动，而且新鲜的空气对我们的健康有好处。
9. You look a bit pale. 你脸色似乎不太好。
10. I hope you get well (better) soon. 我希望你不久就能康复。
11. I'm sure you'll get it over soon. 我相信你不久就会痊愈。
12. I haven't been feeling well recently. 最近我感觉不太好。
13. Take things easy. 别紧张。
14. Have a good rest. 好好休息一下。
15. Go and see a doctor. 去看看医生吧。

莎莉邀请李明去她家吃饭，期间他们聊得很开心。你将听到以下对话，请按要求填充表格：

Items	Information
Food Sally likes	
Li Ming's favorite food	
Food Li Ming suggests	
Ingredients in Shanghai food	
Time to taste the new food	

Useful Expressions 常用句型

1. Dinner is ready. 晚餐准备好了。

2. I didn't know you are such a good cook. 我不知道你是个那么好的厨师。

3. All the dishes look so delicious. 所有的菜看起来那么美味。

4. Just help yourself to anything you like. 爱吃什么自己动手。

5. It is a surprise for me that you know how to make Cantonese food. 你会做广东菜真让我吃惊。

6. I like Cantonese food. It is light and tastes good. 我喜欢广东菜。它比较清淡而且美味。

7. But my favorite is Sichuan food. 但是我最喜欢的是四川菜。

8. Sichuan dishes are delicious, but they are too spicy. 四川菜很好吃,但它们太辣了。

9. It is a bit heavier than Cantonese food, and it uses a lot of seafood and fish. 它比广东菜口味重一点,而且用大量的海鲜和鱼来做菜。

10. I'm fond of seafood. 我喜欢吃海鲜。

11. I know a famous restaurant which serves very good Shanghai food. 我知道一家很出名的餐馆提供很好吃的上海菜。

大卫和安娜在讨论星期日的电视节目。你将听到以下对话,请按要求填充表格:

Items	Information
TV program David watched	
The movie Anna watched was about	
David's favorite movies	
Whose live concert	
Replay time	

Useful Expressions 常用句型

1. What did you watch on TV yesterday? 你昨天看了什么电视节目?

Module 4　English Audio Typing Practice

2. I watched some game shows. 我看了几场比赛。
3. I like to try to answer the questions with the contestants. 我喜欢和参赛者一起回答问题。
4. I watched a movie on Channel Six. 我在第六频道看了一部电影。
5. I prefer the action movies. 我喜欢动作片。
6. Have you watched the live concert of Michael Jackson? 你看迈克的现场演唱会了吗?
7. There will be a TV replay at 8∶00 Tuesday evening on Channel Three. 星期二晚上8点整, 在第三频道, 有电视转播。
8. Is there anything worth watching on the other channel? 另一个频道有什么值得看的节目吗?
9. I think there's a Western on. 我想那个频道在放美国西部片。
10. Do you know what's on after the news? 你知道新闻节目后是什么吗?
11. I believe there's a TV special. 我想是一部电视专题片。
12. What's on Channel Two at 8∶00? 第二频道8点钟是什么节目?
13. There is a sitcom I watch every week. 有一部我每个星期都看的情景喜剧片。

ABC 公司销售代理要出差去美国, 打电话到航空公司预订一张机票。你将听到他和航空公司票务员的对话, 请按要求填充表格:

Items	Information
Flight No. & destination	
For whom	
Departure time	
Ticket type	
Fare	

Useful Expressions 常用句型

1. I'd like to book a ticket to London on January 10th. 我要预订一张1月10日去伦敦的机票。
2. Flight 682 will leave for London at 10∶00 a.m. that day. 航班682会在那一天上午10点整飞往伦敦。
3. There are still some seats available. 现在还有一些位置可供预订。
4. I will book one business class. 我要预订一张商务舱的机票。
5. Now you have been booked. 你已经预订成功。
6. The departure time is 10∶00 a.m. 出发时间是上午10点整。
7. What is the fare? 机票费用多少?

一个中等职业学校的应届毕业生,现在一家公司接受面试,你将听到以下对话,请按要求填充表格:

Items	Information
Interviewee's name	
Major	
Language ability	
Computer skill	
Hobbies	

Useful Expressions 常用句型

1. Please tell me something about your education first. 请先告诉我你的受教育情况。
2. My major is Business English. 我的专业是商务英语。
3. How is your language ability? 你的语言能力怎样?
4. I passed PETS – 2. 我通过了公共英语二级。
5. I am familiar with Microsoft Office software. 我能熟悉操作微软办公软件。
6. We'll inform you soon. 我们会尽快通知你。

ABC 公司销售部的前台接待员接远东公司的盖茨先生来电。你将听到以下对话,请按要求填充表格:

Items	Information
Caller's name	
Caller's number	
To whom	
Message	
Date/time	

Useful Expressions 常用句型

1. I'm Tom Gates from Far-East Corporation. 我是远东公司的汤姆·盖茨。
2. I would like to make an appointment with Mr. Chen this week. 我想和陈先生约这个星期见面。
3. I want to see him about some details of the contract. 我想和他谈一下有关合同的细则问题。

Module 4　English Audio Typing Practice

4. Let me check Mr. Chen's diary. 让我查查陈先生的日程安排。
5. Would 9：30 Tuesday morning be convenient for you? 星期二上午 9：30 分对你是否合适？
6. Would you mind leaving your contact number? 你是否介意留下你的联系电话？

一个人在美国旅游，现在想去唐人街，但不知道该怎么走，需要向一个警察问路。你将听到以下对话，请按要求填充表格：

Items	Information
Place she wants to go to	
Transportation she should take	
Location of subway station	
Station to get off	
Fare	

Useful Expressions 常用句型

1. Can you tell me how to get to Chinatown? 你能告诉我怎么去唐人街吗？
2. It's a long way from here. 去那里路程很远。
3. I think you'd better take the subway. 我想你最好坐地铁去。
4. You should take Subway One to Mayflower Park. 你应该搭乘一号地铁线到五月花公园。
5. Take Subway One and get off at Mayflower Park. 搭乘一号地铁线，在五月花公园站下车。
6. What is the fare? 票价多少钱？

一个中国女孩和美国的汤姆聊天，谈话中提到两国的国庆节。你将听到以下对话，请按要求填充表格：

Items	Information
Tom's National Day	
How to celebrate in America	
How long is Chinese National Day's holiday?	
Lily's program this year	
When is Lily's program?	

Useful Expressions 常用句型

1. We call it Independence Day. 我们把它称为独立日。
2. On that day, there will be parades during the daytime and firework displays in the evening. 在那一天，白天有游行庆典，晚上则有烟花汇演。
3. That sounds great. I can't wait for it. 这听起来棒极了。我都等不及了。

一个销售代表本来是约了一个客户明天见面，但现在他必须参加一个重要的会议。你将听到以下对话，请按要求填充表格：

Items	Information
From whom	
To whom	
Original meeting time	
Reason for not coming	
New meeting time	

Useful Expressions 常用句型

1. I have made an appointment with Mr. Walker at 10：00 tomorrow morning. 我已经和沃克先生约好在明天上午10点整见面。
2. I have to attend an important meeting in New York tomorrow. 我明天必须参加在纽约召开的一个重要会议。
3. Can we make it some other time? 我们可以另外约一个时间吗？
4. Would 9：00 next Tuesday morning be convenient for you? 下星期二上午9点钟你是否方便？

Mary 今天不舒服，部门经理批准她请两天的病假，这时她需要请求她的一位同事在这两天内帮忙完成一些日常工作。你将听到以下对话，请按要求填充表格：

Items	Information
Colleague's name	
The number of copies needed	
Visitor's name	
Daily office work	
Mary's number	

Module 4 English Audio Typing Practice

Useful Expressions 常用句型

1. Would you do me a favor? 你能帮我一个忙吗?
2. Mr. Chen has granted me two days' leave. 陈先生批准我两天的病假。
3. Please make 10 copies of the report for Mr. Chen. 请将这份报告复印十份给陈先生。
4. Please receive him for me. 请帮我接待他。
5. Leave that to me. 交给我吧。
6. If there are any phone calls in, please answer them for me. 如果有任何电话打进来,请帮我接听。

有人到北京出差,晚上入住一家酒店,现在在前台投诉。你将听到以下对话,请按要求填充表格:

Items	Information
Room number	
What does she complain?	
How to handle the problem	
Room service	
Phone number	

Useful Expressions 常用句型

1. I am staying in Room 1605. 我住在 1605 房。
2. I'll send a clerk to handle this matter immediately. 我马上派一个人过去处理这件事。
3. I'll bring in some clean towels together with some hot water. 我会拿一些干净的毛巾和热水过去。
4. Is there anything else? 还有其他需要吗?
5. If there is anything you need, you can dial 83762419. 如果你还需要什么服务,可以打电话 83762419。
6. I'm always at your service. 随时为您服务。

Mr. Grant 搬新家,朋友询问搬家的情况,有关房间、厨房、花园等。你将听到以下对话,请按要求填充表格:

英语听打实训教程

Items	Information
Time for moving into the new house	
Number of children	
Age of the daughter	
Location of the garden	
Things in the garden	

Useful Expressions 常用句型

1. Moving company. Can I help you? 你好，搬家公司。
2. Yeah, I'm going to be moving from Wan Quan Zhuang to Hua Yuan Cun. 你好，我要从万泉庄搬到花园村。
3. OK, when will that be? And your address? 好的，具体时间？您的地址？
4. Apartment 341, Building No. 35, Wan Quan Zhuang. 万泉庄三十五号楼三四一室。
5. I'm calling to ask how much your rates are for movers. 我打电话是要问一下搬运工人价格怎么来定。
6. The price depends on the amount of items to be moved, and which floor they are to be moved to. 价钱要根据搬运多少件东西，要搬到几层来决定。
7. Generally, it would be 150 Yuan to 200 Yuan within the city limits. 一般来说，在市内是 150 元到 200 元之间。
8. I want to move from Xizhimen to Dongzhimen. 我想从西直门搬到东直门。
9. Do you have a lot to move? 你有许多东西要搬吗？
10. No, I have a television, a fridge, clothes and some smaller items. 不，我有一台电视、一个冰箱、一些衣服和一些小物品。

有一个同学住院了，其他同学都很关心他，并询问他的病情。你将听到以下对话，请按要求填充表格：

Items	Information
Who's in hospital?	
What's the matter with Bob?	
Where did he come back?	
Disease	
Present condition	

Useful Expressions 常用句型

1. What seems to be bothering you? Have you got a high fever? 您觉得哪儿不舒服？您发烧吗？

Module 4 English Audio Typing Practice

2. Do you have a record? 您有病历吗？
3. I'll transfer you to the surgery department. 我给您转到外科去。
4. What's wrong with you? 您怎么了？
5. Are you bringing anything up when you cough? 咳嗽时有痰吗？
6. Have you had any chills (chest pain)? 您有发冷（胸痛）吗？
7. Have you ever coughed up blood 您咳血吗？
8. All right. Let me examine you. Would you mind taking off your coat? 好吧。我给您检查一下，您不介意脱掉外衣吧？
9. Take a deep breath, please. 请深呼吸。
10. What kind of treatment have you had? 您过去用什么方法治疗？
11. Where is your pain? 您觉得哪儿痛？
12. What's your appetite? 您的胃口怎么样？
13. How's your appetite these days? 近来食欲怎么样？
14. How long have you been feeling like this? 您有这样感觉多长时间了？
15. Have you been coughing and sneezing? 您是不是一直咳嗽和打喷嚏？

你将听到一段关于天气和天气预报的对话，请按要求填充表格：

Items	Information
What are they going to do?	
Time for the activity	
Weatherman says	
What's the weather like now?	
Temperature this afternoon	

Useful Expressions 常用句型

1. It looks it's going to be sunny. 今天看来像是个晴天。
2. They say we're going to get some rain later. 据说待会儿要下雨。
3. I just hope it stays warm. 希望一直暖和下去。
4. I think it's going to be a nice day. 我想今天会是好天气。
5. It's certainly a big improvement over yesterday. 肯定比昨天大有好转。
6. But it's supposed to get cloudy and windy again this afternoon. 但是，据说今天下午又要转阴刮风了。
7. Well, the worst of the winter should be over. 不过，冬天最糟糕的一段日子总该过去了。
8. It seems to be clearing up. 看来天要放晴了。
9. It's such a nice change. 真是令人高兴的转变。
10. I really don't think this nice weather will last. 我确实认为这样的好天气持续不了多久。

11. Let's just hope it doesn't get cold again. 但愿不会再冷。
12. As long as it doesn't snow! 只要不下雪就行啊!

17

老板称赞一个员工的工作出色,员工谦虚地回答,老板表示赞赏和鼓励。你将听到以下对话,请按要求填充表格:

Items	Information
Clerk's name	
Boss's name	
Result of her annual report	
Boss's comment on the clerk	
Boss's suggestion	

Useful Expressions 常用句型

1. Well done. 干得不错。
2. You speak very good English. 你的英语说得真好。
3. You are great. 你真棒。
4. You have done a good job. 你表现不错。
5. Thank you. It's nice of you to say so. 谢谢,您这样说太客气。
6. Thank you. Yours is even nicer. 谢谢,你的还更好呢!
7. It's my great fortune to work with you all. 能与你们一同工作,我真是幸运之极。
8. I am all ears. 我洗耳恭听。
9. I'd like to have your views on this issue. 我想知道您对此事有何高见。
10. Thank you, but it really isn't anything special. 谢谢,不过这实在是不值一提。
11. I'm honored, but I'm not sure I'm the right person. 很荣幸,不过我是否胜任,可没把握。

18

大卫的父亲打电话给他的老师,讨论大卫学习的事情。你将听到以下对话,请按要求填充表格:

Items	Information
Name of David's father	
Teacher's name	
David's present condition	
Father's purpose for calling	
Time for David to go back to school	

Module 4　English Audio Typing Practice

Useful Expressions 常用句型

1. Thanks a lot. 多谢。
2. Thank you for your help. 谢谢您的帮助。
3. Thank you. It was very kind of you. 谢谢，您太好了。
4. I'm very much obliged to you. 非常感谢您。
5. I don't know how I can thank you enough. 我不知道怎样谢您才好。
6. You've done me a great favor. 您帮了我大忙。
7. I don't know what I would have done without your help. 要不是您的帮助，当时我真不知道该怎么办。
8. You're welcome. 别客气。/Not at all. 没什么。/ Don't mention it. 不必谢。
9. It's a real pleasure for me to do it. 我很乐意这样做。
10. No trouble at all. 一点儿也不麻烦。

学生安德鲁遇见老师苏珊，老师给了安德鲁一些学习英语的建议。你将听到以下对话，请按要求填充表格：

Items	Information
How did Andrew look today?	
Mark/grade in the exam	
How hard is he learning English?	
What should he do if he wants to learn English well?	
What should he do to improve his sense of language?	

Useful Expressions 常用句型

1. Now you've got it. 现在你做到了。
2. You're incredible! 你简直难以置信！
3. You can do it. 你能做到。
4. How did you do that? 你怎样完成的？
5. You're fantastic! 你真是太妙了！
6. You're improving. 你在进步。
7. You're on target. 你达到目标了。
8. You're on your way. 你在前进中。
9. Good job! 干得出色！
10. That's incredible! 简直难以置信！
11. Let's try again. 再试试。
12. You're unique. 你太不寻常了。

13. Nothing can stop you now. 现在你已所向无敌。
14. You're a winner! 你是赢家！
15. You'll make it. 你一定会成功的。

20

父亲责备女儿看电视过久，而女儿也要求父亲减少抽烟，父女俩最后达成协议。你将听到以下对话，请按要求填充表格：

Items	Information
How long has Nicole been watching TV today?	
What day is it today?	
What should Nicole learn to protect ?	
How many packages of cigarettes does the father smoke a day?	
Dad's promise	

21

丹尼尔不小心把萨拉的书弄脏了，并向她道歉。你将听到以下对话，请按要求填充表格：

Items	Information
What happened to the book?	
Weather at the time	
Name of the book	
When was the book borrowed?	
How to compensate	

22

有人想去颐和园，但是不知道怎么走，去问路，路人告诉他如何去那里。你将听到以下对话，请按要求填充表格：

Items	Information
Destination	
How far is it?	
Which turn?	
To the right or left?	
Is it far?	

Module 4   English Audio Typing Practice

Useful Expressions 常用句型

1. No, but I can ask the way. Excuse me, can you tell me where the Prince's Building is? 不认识，但我可以问路。对不起，你能告诉我王子大厦在哪里吗？
2. I'm sorry. I'm a stranger here myself. 很抱歉，我也不熟悉这里。
3. Excuse me. How do I get to the Prince's Building, please? 打扰了，请问王子大厦怎么走？
4. Well, turn to the left at the first corner after the crossroads. It's near the corner. You can't go wrong. 噢，过十字路口后在第一个拐角向左拐，离拐角不远。你会找到的。
5. Is it far from here 离这儿远吗？
6. No, it's only a couple of blocks away. 不远，过两个街区就到了。
7. Don't mention it. 不客气。
8. Can you tell me the way to the bus station? 请问去汽车站怎么走？
9. Go straight ahead and turn left at the traffic lights. 沿这条路一直向前走，在红绿灯那儿向左转。
10. Would you please tell me where the post office is? 请问邮局在哪儿？
11. Would you please tell me if there is a hospital nearby? 请问附近有医院吗？
12. Is the zoo far from here? 动物园离这儿远吗？
13. Will it take long to get to the airport? 去机场要很长时间吗？
14. Go along the street until you come to the traffic lights. 沿这条路一直走到红绿灯那儿。
15. Turn right/left at the second crossing. (Take the second turning on the right / left.) 在第二个十字路口向右/左转弯。
16. Take a number 46 bus, and get off at the Square. 坐46路公共汽车，在广场下车。
17. It's at the corner of Huaihai Street and Xizang Road. 在淮海街和西藏路的路口。

卡伦走进专卖流行衣服的小商店。你将听到以下对话，请按要求填充表格：

Items	Information
What to buy	
Material wanted	
For which season	
Price	
Discount	

Useful Expressions 常用句型

接待:

1. Did you find anything you like? 你找到你喜欢的吗？

2. What can I do for you? 你要些什么?

3. Can I help you? 我能帮你吗?(需要些什么?)

选择与购买:

1. I want a pair of shoes/a jacket. 我想买一双鞋/一件夹克。

2. I'd like to see some towels. 我想看看毛巾。

3. Show me that one, please. 请把那个给我看看。

4. Would you show me this cup? 你能把这只杯子让我看一下吗?

5. I'm interested in this new model of car. 我对这款新车有兴趣。

6. I'm just looking, thanks. 我只是看看,谢谢。

试穿:

1. Could you try it on, please? How is it? 请试穿看看好吗? 如何?

2. I like this one. May I try it on? 我喜欢这一种。我能试穿吗?

询问:

1. Do you have any on sale? 你们有什么特卖品吗?

2. Do you carry a hundred percent cotton pants? 你们有百分之百纯棉的裤子吗?

3. If I ordered a suit now, how long could it take before I got delivery? 如果我现在订一件西装,要多久才能接到货?

尺寸和颜色:

1. The fit isn't good. 尺寸不太合适。

2. It's too big. /Too small. 太大了。/太小。

3. It seems to fit well. 好像蛮合身的。

4. Can I have a size larger? 可以给我一个大一点儿的吗?

5. How about this blue one? 这个蓝色的怎么样?

价格:

1. How much does it cost? 多少钱?

2. What's the price of this suit? 这套西装多少钱?

3. How much do I have to pay for it? 我要付多少钱?

4. How much are these ties? 这些领带要多少钱?

5. I'll give it to you for US $ 5250. 5 250美金卖给你。

6. Can you make it cheaper? 你能便宜点吗?

付钱:

1. How can I pay for it? 我要如何付钱?

2. May I write a check for you? 我能开支票吗?

3. Do you take traveler's checks? 你们接受旅行支票吗?

4. Sorry, we don't take checks. 对不起,我们不接受支票。

王女士在医院看病,医生对她进行检查。你将听到以下对话,请按要求填充表格:

Module 4 English Audio Typing Practice

Items	Information
Mrs. Wang's problem	
How well has Mrs. Wang slept?	
What Mrs. does the doctor do then?	
Has Mrs. Wang had a cold?	
Doctor's suggestion	

Useful Expressions 常用句型

1. Are you feeling Okay? 你感觉可好？
2. I'm not feeling well. 我觉得身体有点儿不对劲。
3. What are your symptoms? 是什么症状？
4. I feel like throwing up. /I feel sick. 我想吐。
5. Did you eat anything unusual? 吃了什么不对劲的东西没有？
6. Do you have a fever? 你发烧吗？
7. Let me check your temperature. 量一下体温吧。
8. He has a headache/stomach ache/He feels light-headed. 他觉得头痛/胃痛/头晕。
9. She has been shut-in for a few days. 她生病在家几天了。
10. He has a headache, aching bones and joints. 他头痛，骨头、关节也痛。
11. He has a persistent cough. 他不停地在咳。
12. He is sleeping poorly. 他睡不好
13. He has to breathe through his mouth. 他要用口呼吸。
14. He has an uncomfortable feeling after meals. 他饭后肚子觉得胀胀的，很不舒服。
15. His blood pressure is really high. 他的血压很高。

25

弗兰克和格林在谈论各自的爱好。你将听到以下对话，请按要求填充表格：

Items	Information
Frank's fond of _____	
Frank's keen on _____	
Green's hobby	
Frank's interested in _____	
Who collects stamps?	

Useful Expressions 常用句型

1. Do you have any hobbies? 你有什么爱好吗？

2. What are you interested in? 你对什么比较感兴趣?
3. What are your interests? 你的爱好是什么?
4. What do you do in your spare time? 空闲时间你干什么?
5. How do you spend your holidays? 假期你都是怎么过的?
6. Lots of people like stamp collecting. 许多人喜欢集邮。
7. I'm a Michael Jordan's fan. 我是迈克尔·乔丹迷。
8. Does his film appeal to you? 你喜欢他的电影吗?
9. Do you go for picnics? 你常去郊游吗?
10. I particularly like English literature. 我对英国文学情有独钟。
11. What's so interesting about football? We girls don't like it. 足球有什么意思? 我们女生不喜欢它。
12. She has a particular interest in painting. 她特别爱好绘画。
13. I often take my mind off my work by reading an interesting novel. 通常我通过阅读小说把注意力从工作上转移过来。
14. He plays the violin just for enjoyment. 他拉小提琴只是为了自娱自乐。
15. Photography is an expensive hobby. 摄影是门花费很多的爱好。

今天 Ellen 上了第一堂英语课,老师告诉他们有三种方法可帮助他们学习语言。你将听到以下对话,请按要求填充表格:

Items	Information
What class?	
How was it?	
First suggestion	
Second suggestion	
Who to imitate	

Useful Expressions 常用句型

1. Do you speak English? 你会说英语吗? Yes, a little. 会讲一点儿。
2. How long have you learned English? 你学英语多久了?
3. He speaks English fluently. 他讲英语很流利。
4. Your English is very good. 你的英语很好。
5. You speak English pretty well. 你的英语讲得很好。
6. Are you a native speaker of English? 你的母语是英语吗?
7. My native language is Chinese. 我的母语是汉语。
8. He speaks with a London accent. 他带点伦敦口音。

Module 4　English Audio Typing Practice

9. He has a strong accent. 他口音很重。
10. I have some difficulty in expressing myself. 我表达起来有点困难。
11. I'm always confused with "s" and "th". 我常把"s"和"th"搞混。
12. Can you write in English? 你能用英文写文章吗?
13. Your pronunciation is excellent. 你的发音很好。
14. How can I improve my spoken English? 我该怎样才能提高口语水平?
15. Listen and repeat. 先听,然后再重复一遍。

一位顾客刚到饭店,正在柜台登记。你将听到以下对话,请按要求填充表格:

Items	Information
What's the customer's name?	
What is the room number?	
How many days?	
How much a day?	
What is the check-out time?	

Useful Expressions 常用句型

1. I'd like to check in, please. 你好,我想入住贵酒店。
2. Certainly. Can I have your name, please? 当然可以,能把姓名告诉我吗?
3. Will you need a wake-up call, sir? 先生您需要唤醒服务吗?
4. Yes, please. At 6:30 a.m. 是的,请在早上6:30唤醒我。
5. OK, then, your room is 502 on the fifth floor. Breakfast is served between 6:30 and 9:00 a.m. Enjoy your stay. 好的。您的房间号码是五层的502房间。早餐时间是6:30到9:00。祝您入住愉快。
6. Good evening, I have a reservation under the name of Tomlinson. 晚上好,我有预订,名字是Tomlinson.
7. OK, I've found it. Checking out on the 27th? 好,我找到了。是27日退房吗?
8. Can I use a credit card for the deposit? 我能用您的信用卡划账押金吗?
9. Yes, sure. Also, I'd like a non-smoking room, please. 当然可以。另外,请给我一间无烟房间。
10. Certainly, madam. Here's your key. Your room is on the 7th floor and on the left. Room 781. Check-out is at 12:00. 当然可以,女士。这是您的房间钥匙。您的房间在7层左侧,房间号码781。退房需要在12点之前。
11. How long will you be staying? 您打算住多久?
12. You have altogether four pieces of baggage? 您一共带了4件行李,是不是?

英语听打实训教程

13. Excuse me, where can I buy some cigarettes? 劳驾，我到哪儿可买到香烟？
14. We have a Chinese restaurant and a Western-style restaurant. Which one do you prefer? 我们有中餐厅和西餐厅，你愿意去哪个？
15. I'd like to try some Chinese food today. 今天我想尝尝中国菜。

一个女人正在咨询关于当地博物馆和美术馆的信息。你将听到以下对话，请按要求填充表格：

Items	Information
Open time	
How much?	
How many museums?	
Names of the museums	
Name of the gallery	

Useful Expressions 常用句型

1. Would you like to visit our factory some time? 什么时候来看看我们的工厂吧？
2. I can set up a tour next week. 我可以安排在下个礼拜参观。
3. The tour should last about an hour and a half. 这次参观大概需要一个半小时。
4. I'm really looking forward to this. 我期待这次参观很久了。
5. I plan to get to your showroom. 我打算到你们的展示中心看看。
6. I'll meet you there; shall we say about eleven o'clock? 我会在那儿等你，你看十一点左右如何？

一位男士正在向一位女士问路。你将听到以下对话，请按要求填充表格：

Items	Information
What to buy	
Where to buy	
Number of the bus	
Place to take a bus	
Place to get off	

Module 4　English Audio Typing Practice

Useful Expressions 常用句型

1. 问路时常用的方位词：east 东、south 南、west 西、north 北、left 左、right 右、straight on 往前直去、there 那儿、front 前方、back 后方、side 侧旁、before 之前、after 之后、first left/right 第一个转左/右的路

2. 问路：

   How can I get to the...? 请问如何前往……？

   Where is the nearest...? Is there... nearby? 请问附近有没有……？

3. 提供帮助：Can I help you?

4. 问距离：How far away is the nearest...?

30

一位男士被安排去机场接机，你将听到以下对话，请按要求填充表格：

Items	Information
Who will meet Sally?	
Sally's appearance	
Time for Sally's arrival	
Sally's coming from	
Sally's flight number	

Useful Expressions 常用句型

1. 请求帮助：

   Will you do me a favor?

   Can you help me?

   Will you give me a hand? Would you...?

   Could you...? Would you mind doing...?

2. 表示愿意帮忙：Yes, of course. /Sure. /No problem. /All right.

3. 表示不能提供帮忙：I'm afraid I can't..., because... 恐怕不行，因为……

   I'd like to help, but... 我很乐意帮忙，但……

4. 表示对别人帮助的感谢：

   Thank you. /Thanks a lot.

   I really appreciate it.

   It's very kind of you...

   How can I ever thank you!

   You are so kind to take the trouble to help me.

   If there is anything else I can do, please let me know.

31

一位男士打电话到电影院,你将听到以下对话,请按要求填充表格:

Items	Information
Purpose of calling	
Date for the concert	
Number of the tickets	
Telephone number	
Where and when to get the tickets	

Useful Expressions 常用句型

订票基本句型:

1. I'd like to book 4 tickets, please. 我订 4 张票。
2. Would you like one way or round trip? 你要单程还是往返票?
3. Round-trip. We'll return on… 往返票。我们将在……返回。
4. tickets on… to… and returning to… on… ……(日)到……(地点),……(日)返回到……(地点)的票

订票询问价格:

1. Could you tell me how much it costs to…? 你能告诉我……多少钱吗?
2. The price of a ticket from… to… is… 从……到……的票价是……。

一位乘客在机场的入境处和行李处,工作人员正在检查他的行李。你将听到以下对话,请按要求填充表格:

Items	Information
What to see	
How long will he stay?	
Purpose for visit	
Things to declare	
What kind of form?	

Useful Expressions 常用句型

1. Which gate does my flight leave from? 请问我所乘坐航班的登机口是哪一个?
2. Is this the way to Gate 9? 这里是去 9 号登机口吗?

Module 4　English Audio Typing Practice

3. Where can I get my luggage? 请问去哪儿取行李？
4. Here is my claim tag. 这是我的行李单.
5. I'd like to check these bags! 我要托运这些行李。
6. Shall I reserve you a seat, then? 我能为你预订座位吗？
7. Please put the luggages here for a security check. 请把行李放在这里安检。
8. Would you like an economy seat or a first class seat? 你想要经济舱还是头等舱？
9. I'm sorry but our flights are fully booked on that day. 很抱歉那天机票已订完，没有空位了。
10. Do you want a seat in the smoking or non-smoking section? 你想坐在吸烟区还是非吸烟区？

33

一位客人正在餐馆点餐，服务员在帮助他。你将听到以下对话，请按要求填充表格：

Items	Information
Kind of beef	
Vegetable ordered	
Drink ordered	
Things on the table	
Dessert ordered	

Useful Expressions 常用句型

1. Order a meal 点菜
2. Are you ready to order yet, sir? 先生，可以点菜了吗？
3. May I take your order? 您点点儿什么？
4. What would you like? 你想吃点什么？
5. Anything else? 还要别的吗？
6. Would you like some coffee? 来杯咖啡怎么样？
7. May I see your menu, please? 能给我看看菜单吗？
8. I'd like to see a menu, please. 请给我菜单。
9. May I see the wine list, please? 请给我看一下酒单，好吗？
10. What do you recommend? 有什么菜可以推荐的吗？
11. What is your suggestion? 你有没有什么好介绍？
12. Do you have any local specialties? 您这儿有什么地方风味吗？
13. I'd like a cup of coffee, please. 请给我一杯咖啡。

一位乘客在飞机场，正在向工作人员进行咨询，你将听到以下对话，请按要求填充表格：

Items	Information
Purpose of asking	
The way of flying	
How many airlines to choose from	
Price of C. A. A. C.'s return ticket	
Price of B. A.'s one-way ticket	

Useful Expressions 常用句型

1. What flights do you have from Beijing to Hong Kong every week? 从北京到香港每周有哪些航班？

2. I'd like to book a flight to Hong Kong on Tuesday morning. 我想订一张星期二早上去香港的机票。

3. What's the price for a return trip? 回程机票多少钱？

4. What is the price for a one-way trip? 单程票多少钱？

5. Do you want to fly direct, or do you want to stop over somewhere on the way? 你想直飞还是中途转机？

6. How long does it take to fly from Beijing to Hong Kong? 从北京到香港要飞多长时间？

7. Is the flight delayed? 飞机晚点了吗？

8. At what time will it land? 飞机什么时候降落？

9. Can I see your passport and your ticket, please? 我能看一下你的护照和机票吗？

10. Do you have anything to declare? 你有什么东西要申报吗？

玛丽刚刚买了一台新电脑，但遇到一些问题。你将听到以下对话，请按要求填充表格：

Items	Information
Owner of the computer	
When was the computer bought?	
Price of the computer	
Problem with the computer	
Solution to the problem	

Module 4　English Audio Typing Practice

Useful Expressions 常用句型

1. a modem and an ADSL adapter 调制解调器或 ADSL 适配器
2. sharing disks 共享磁盘
3. Networks are about three things: exchanging files, sharing resources and running common programs. 网络涉及三件事情：交换文件、共享资源及运行公用程序。
4. sharing resources 共享资源
5. The files can be transmitted by network cable. 文件可以通过网线传送。
6. log in/on 登录
7. a LOGIN command 登录命令
8. password 密码

一位男士和一位女士正在谈论别人结婚的事情，你将听到以下对话，请按要求填充表格：

Items	Information
What's the surprising news?	
Lucky guy's name	
Time to get married	
Maid's name	
Best man's name	

Useful Expressions 常用句型

1. wedding ceremony 结婚典礼
2. say one's vows 立下婚誓
3. wedding day 举行婚礼的日子
4. wedding anniversary 结婚周年纪念日
5. bride 新娘
6. bridegroom or groom 新郎
7. best man 伴郎
8. maid of honor 伴娘
9. honeymoon 蜜月
10. wedding dress 婚纱、结婚礼服
11. wed in a civil ceremony 登记结婚
12. marriage certificate 结婚证

英语听打实训教程

爱丽丝等人假期准备去旅游，正在就有关事项进行讨论。你将听到以下对话，请按要求填充表格：

Items	Information
Where to go	
When to set out	
How long to get there	
How to get there	
What kind of clothes to take	

Useful Expressions 常用句型

1. This is the best season to go traveling/sightseeing. 这是去游玩的最好时节。

2. We're going to the Great Wall. 我们打算去长城。

3. We'll meet at the school gate at 8 o'clock tomorrow morning. 明天早上 8：00 我们将在学校门口碰面。

4. How do we get there? 怎么去那里呢？

5. Who will you go with? 你同谁一起去呢？

6. Beijing deeply impresses me. 北京给我留下了深刻的印象。

7. It's my great honor to have this chance to visit London. 有这样的机会参观伦敦是我最大的荣幸。

8. It's true that "seeing is believing". "眼见为实"是真理。

9. Oh, it's too late. I must go now. Thank you for your guidance. 太晚了，我得马上走了。谢谢你给我们做向导。

10. How long will it take for me to visit the city? 参观这个城市要多长时间？

鲍勃在街上遇见好久不见的朋友，并跟她聊起自己的工作。你将听到以下对话，请按要求填充表格：

Items	Information
Bob's present job	
Bob is thinking of _____	
What kind of job does Bob like?	
Bob's dislikes	
Bob's ideal jobs	

Module 4 English Audio Typing Practice

Useful Expressions 常用句型

1. Is your work very hard? 你的工作难度大吗?
2. The work is hard, but somewhat challenging. 这份工作不轻松,但极具挑战性。
3. That's a job that pays well. 这是一份薪水不错的工作。
4. Where do you work now? 你现在在哪里工作?
5. What kind of work do you do? 你是做什么的?
6. I'm a clerk. 我是名职员。
7. It's difficult to get along with some colleagues. 同事之间很难相处。
8. It's a job with no paid holidays. 这份工作没有带薪假期。
9. The work is endless. 工作没完没了。
10. May I ask why you want to change your job? 我能问一下你换工作的原因吗?

ABC 公司的面试官正在对杰克进行面试,你将听到以下对话,请按要求填充表格:

Items	Information
Interviewee's age	
University that Jack graduated from	
Interviewee's English level	
Expected salary	
Time to get the result	

Useful Expressions 常用句型

1. What kind of job are you looking for? 你想找什么样的工作?
2. Could you tell me something about yourself? 你能跟我说说你的情况吗?
3. Would you mind answering a few personal questions? 你介意回答一些私人问题吗?
4. Why do you want to work here? 你为什么要到我们这里工作?
5. Which school did you graduate from? 你毕业于什么学校?
6. What's your major? 你学什么专业?
7. Are you single or married? 你未婚还是已婚?
8. What are you good at? 你擅长什么?
9. Can you type? 你会打字吗?
10. How much do you expect per month? 你希望月薪多少?

两位同班同学正在讨论毕业以后如何计划自己的未来,你将听到以下对话,请按要求填充表格：

Items	Information
What does Peter want to do after graduation?	
What's Mary's parents' advice?	
What does Mary want to do?	
Mary will regret	
Peter's hope for Mary	

Useful Expressions 常用句型

1. I'd like to go into management. 我想进入管理层。
2. I hope I can finally enter the university. 我希望我最后能进入大学。
3. What do you hope to do when you finish school? 毕业以后你希望做什么？
4. I knew it would be impossible for me to be successful in that field. 我知道对我来说要在那个领域取得成功是不可能的。
5. As long as you plan carefully, everything is possible. 只要你细心准备，一切都是可能的。
6. I am looking forward to my new life. 我正期待着我的新生活。
7. It must be interesting to live outside the campus alone, I guess. 我想独自一人生活在校外是有趣的。
8. Where are you going after you get married? 你结婚以后去哪里？
9. I won't give up my job at present since I enjoy it and still need it to support my family. 目前我不会放弃工作，因为喜欢我的工作，并且仍旧需要工作养家糊口。
10. We're going to buy a flat or a small house somewhere downtown. 我们会在闹市区的某个地方买一套公寓或者一座小房子。

杨红来到机场接一个外宾的飞机。你将听到以下对话，请按要求填充表格：

Items	Information
Foreigner's name	
Name of the woman's Company	
Where are they?	
Where to take a rest	
Where to go afterwards	

Module 4 English Audio Typing Practice

Useful Expressions 常用句型

初次见面的自我介绍

1. May I introduce myself? 我可以自我介绍一下吗?
2. Hello, I'm Tom. 你好，我叫汤姆。
3. Excuse me, I don't think we have met. My name is Tom. 不好意思，我们还没见过面，我叫汤姆。
4. How do you do? I'm Tom. 你好! 我叫汤姆。
5. First, let me introduce myself. I'm Tom, I'm the Production Manager. 首先，我自我介绍一下。我叫汤姆，是生产部经理。

介绍后的回答

1. I am glad to meet you. 很高兴认识你。
2. Nice to meet you. 很高兴认识你。（即将分手时可以说）Nice meeting you.
3. How nice to meet you. 很高兴认识你。
4. I have heard so much about you. 久闻大名。
5. Someone has told me all about you. 久仰大名。
6. I have been wanting to meet you for some time. 久仰久仰。
7. I am delighted to make your acquaintance. 很高兴认识你。
8. It is a privilege to know you. 认识你是我的荣幸。
9. It is a pleasure to know you. 认识你是我的荣幸。

一位客房住客正在给酒店前台服务员打电话，你将听到以下对话，请按要求填充表格：

Items	Information
Purpose of calling	
Time the call is wanted	
Caller's room number	
Caller's name	
Reason for getting up early	

Useful Expressions 常用句型

1. Welcome to our hotel. 欢迎到我们酒店来。
2. I hope you will enjoy your stay in our hotel. 希望您在我们酒店过得愉快。
3. Housekeeping. May I come in? 我是客房服务员，可以进来吗?
4. Can you tell me your room number? 您能告诉我您的房间号码吗?

95

5. May I see your room card? 能看一下您的房卡吗?
6. I'll be with you as soon as possible. 我尽快来为您服务。
7. When would you like me to clean your room, sir? 您要我什么时间来给你打扫房间呢，先生?
8. It's growing dark. Would you like me to draw the curtains for you? 天黑了，要不要我拉上窗帘?
9. We'll come and clean the room immediately. 我们马上就来打扫您的房间。
10. I'm coming to change the sheets and pillowcases. 我来换床单和枕套。

 43

一位顾客来到商场投诉新买的货物，店员正在跟他讨论解决的办法。你将听到以下对话，请按要求填充表格：

Items	Information
What did the man buy?	
What's the problem?	
What did he hear when he switched the MP3 player on?	
How will the woman solve the problem?	
What did the woman ask for?	

Useful Expressions 常用句型

1. May I help you? 有什么事吗?
2. We have a damaged shipment from you. 你们送来的货有损坏。
3. We'll look into it right away for you. 我们会立刻调查清楚。
4. Just whose fault is this damage? 这次的损坏究竟是谁的责任呢?
5. The goods were in good shape when they left our store. 货离开商店时都是完好无缺的啊。
6. It certainly didn't arrive here that way! 送到这儿时可不是那样!
7. Here is the final settlement for your claim. 你的赔偿问题终于解决了。
8. Thanks, we appreciate the fast work. 谢谢你们这么快就办好了。
9. We only hope we won't have this kind of problem again. 我们仅希望不会再有这样的事情发生。

 44

李先生和爱丽丝在讨论何时面谈机构协议的问题。你将听到以下对话，请按要求填充表格：

Module 4 English Audio Typing Practice

Items	Information
Caller's name	
Purpose of calling	
Will Lee be free in the afternoon?	
When will they meet?	
Where to meet	

Useful Expressions 常用句型

1. I wonder if you are free this afternoon. 你今天下午有空吗?

2. I would like to make an appointment with Miss Li. 我想跟李小姐预约见面。

3. What time would be convenient for you? 您什么时间方便呢?

4. How about Friday morning? 星期五早上怎么样?

5. May I expect you at 10 o'clock? 10 点钟可以吗?

6. Friday would be best for me. 最好是星期五。

7. Sorry, I've got two appointments from two to six. 很抱歉,我从两点到六点有两个预约了。

8. I'm afraid this afternoon I'm fully engaged. 恐怕我整个下午都没有时间。

9. I'm sorry. I won't be here next week. I am going to Beijing on business. 很抱歉,下周我不在,我要去北京出差。

女孩向男孩询问关于网上购物的问题,你将听到以下对话,请按要求填充表格:

Items	Information
Things bought on net	
Reason for buying online	
What does the woman want to buy?	
Who is Jack?	
Woman's second question	

Useful Expressions 常用句型

1. Online shopping has become a new way of shopping. 网上购物是一种新兴的购物方式。

2. There are other ways of shopping nowadays: shopping by mail, shopping by phone, TV shopping and direct shopping. 当今,还有其他的一些购物方式:邮购、电话购物、电视购物和直销。

3. Mary is crazy about online shopping. 玛丽非常热衷于网上购物。
4. I saw a necklace on your website. 我看中你们网站上的一条项链。
5. Do you know the sales number? 你知道销售代码吗?
6. Credit card payment is required. 最好用信用卡付款吧。
7. Do you charge for the delivery? 送货要收钱吗?
8. We usually charge 10% for each delivery, but if the total is more than 100 dollars, we offer a free delivery. 我们的运费一般收货物价格的 10%。但如果买满 100 元以上,我们可以免运费。

Copy Version and Key to Each Exercise

Unit 2

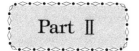

Practice Exercise

Task

Exercise 1
comma, dash— question mark? colon: full stop. semicolon; open quotation mark " close quotation mark" hyphen-open parentheses (close parentheses)

Exercise 2
twenty-five; forty-two; sixty-six; thirty-seven; seventy-nine; eighty-one; one hundred and thirty-two; two hundred and ninety-eight; nine hundred and seven; five hundred and forty-six; three hundred and nineteen; six hundred and thirty-eight; one thousand, two hundred and thirty-eight; one thousand, six hundred and eighty-four; nine thousand, eight hundred and sixty-five; four thousand, five hundred; two million, one hundred and twenty thousand, four hundred and fifty; four million, six hundred and fifty-three thousand, seven hundred and ninety

25; 42; 66; 37; 79; 81; 132; 298; 907; 546; 319; 638; 1,238; 1,684; 9,865; 4,500; 2,120,450; 4,653,790

Exercise 3
88; 53; 67; 72; 91; 98; 299; 361; 432; 558; 813; 940; 3,338; 2,675; 8,932; 5,380; 3,816,780; 6,522,840

eighty-eight; fifty-three; sixty-seven; seventy-two; ninety-one; ninety-eight; two hundred and ninety-nine; three hundred and sixty-one; four hundred and thirty-two; five hundred and fifty-eight; eight hundred and thirteen; nine hundred and forty; three thousand, three hundred and thirty-eight; two thousand, six hundred and seventy-five; eight thousand, nine hundred and thirty-two; five thousand, three hundred and eighty; three million, eight hundred and sixteen thousand, seven hundred and eighty; six million, five hundred and twenty-two thousand, eight hundred and forty

Exercise 1

ability aboard absent accident accountant activity automatic bathroom beautiful battery boring Canada camera certificate chairman childhood Christmas chocolate community company describe dictionary disappointed document employee engineer economy European failure festival

Exercise 2

financial forecast foreigner generation gentleman guarantee headache honesty housewife important impossible independent instruction international journey knowledge language lawyer library location machine manager market manufacture medicine national newspaper normal official operation

Exercise 3

organization passenger performance persuade pleasure pollution powerful presentation production progress purpose quality receipt register restaurant responsible secretary schedule scientific signature situation special statement supermarket tomorrow traditional traffic valuable vocation wonderful

Exercise 1

1. I am going to a concert this weekend.
2. John is a very clever boy.
3. You have really done a good job this time.
4. Only two students failed the final examination.
5. Students should pay attention to what the teacher says in class.
6. The manager is at a meeting and can not answer the call.
7. There was a serious accident around that corner this morning.
8. Jenny is not going to be Tony's partner at tonight's party.
9. I usually have my lunch at the MacDonald's, but the food there isn't healthy.
10. Mary, Susan and I went to the same high school together for three years.

Exercise 2

1. All over the world, people watch soap operas on television.
2. Credit cards are very useful in our daily life.
3. There is a new shopping centre near my home.

Copy Version and Key to Each Exercise

4. Mr. Smith's goods should be sent as soon as possible.
5. Please leave a message for Mary Kate that William Johnson called.
6. I would like to book two plane tickets to Beijing next Friday.
7. Jack is a hard-working person and gets up very early in the morning.
8. Judy is responsible for receiving telephone calls in the company.
9. We find your goods are not up to the international standard.
10. I want to order 4,000 pairs of shoes from your company.

Exercise 3

1. After working for nearly 14 hours, I felt very tired.
2. Yang Hong is now at the airport to meet Mr. Brown.
3. Eye contact is very important in communication.
4. When in Rome, do as the Romans do.
5. Different countries have different cultures.
6. Simon is on a business trip to Shanghai.
7. Helen has been working in that foreign company for 10 years.
8. Studying for a test is important if you want to get a high score.
9. Write down the key points in the discussion and repeat them.
10. We have purchased a coffee-machine from a web store.

Exercise 1

1. The teacher was not happy when she found most of the students didn't do the homework.
2. When I installed the coffee-machine in my house, it exploded and destroyed my table and chairs.
3. I spoke to your service manager last week, and he promised to solve the problem.
4. I have to return this coffee-machine, and I want you to pay $ 2,000 for my damaged furniture.
5. Mr. Johnson is phoning Li Ming to check the arrangements for his visit to the company next Thursday.
6. Nowadays, people are more and more dependent on machines, computers, telephones and the Internet to do business.
7. One evening, William walked into the shopping centre of the hotel and wanted to buy a hat.
8. Fortune Supermarket sells a few things at a lower or special price every week, but this doesn't mean all the prices are lower.

Exercise 2

1. Be careful in the supermarket. You may go home with a bag of food you were not planning to buy.

2. Mr. Lee has been waiting for nearly an hour for his meal, and he is making a complaint to the waiter.

3. The children were always clean no matter how old and shabby their clothing was.

4. At the end of the school year, our teacher moved to another school in the city where most children come from richer families.

5. Susan retired after working for over thirty years in that school and has begun spending her time reading and writing.

6. Martin has been our top salesman and I think he deserves the position of sales manager.

7. I would like to book two double rooms in your hotel for January 1st and 2nd.

8. Nowadays, more people travel by car than any other means of transportation.

Exercise 3

1. I just got a phone call which said that our company's water supply will be turned off tomorrow morning.

2. Mr. Wang reported that the company's production problems have already been solved.

3. This company doesn't treat men and women differently—they employ and pay both equally.

4. Great changes have taken place in China since the policy of opening up and reform.

5. With hard work, she became a great popular writer in her country.

6. I am afraid I have eaten more than enough tonight and I feel uncomfortable in my stomach now.

7. People throughout the world want peace rather than war.

8. To everyone's disappointment, the manager's secretary came to the meeting instead of the manager himself.

Test

1. If you need to leave the dining table to go to the bathroom, say, "Excuse me for a moment, please."

2. Mr. Smith is having an interview for a job as a salesman and he feels very confident in himself.

3. The office secretary handles tasks like typing letters, receiving visitors and answering phone calls.

4. To be successful in a job interview, you should leave a good image of yourself in a short period of time.

5. He told me that he had saved a lot of money by planning first on how to spend it.

6. Your products are always popular among the young with their high quality and fashionable design.

7. If the price can be brought down by 20%, I believe you can double your sales.

8. The teacher wrote out a list of books and assigned the students to read them after class.

Copy Version and Key to Each Exercise

Unit 3

Practice Exercise

Task

Exercise 1

Spring Festival is one of the most important traditional festivals in China. The Chinese people used to celebrate it by pasting festival couplets on the doors, hanging up the red lanterns, and making dumplings. Also, people often called on their relatives and friends, giving them their best wishes.

Exercise 2

Our school is clean and beautiful. There are many trees and flowers around it. In order to make our school more and more beautiful, we should try our best to keep it clean. Don't throw rubbish on the ground. We are responsible to keep our school clean.

Exercise 3

In Chinese culture, colors are given symbolic meanings often different from those in European cultures. The color red always means good luck and prosperity. Gold is the imperial color. White is the color of death (and is the color traditionally seen at funerals). Black symbolizes misfortune.

Exercise 4

Traveling is a common way for Americans to spend their holidays. Also, there are many parks across the United States, among which most of the city parks are free. For example, the famous Central Park in New York is free and it has no walls or fences.

Exercise 1

There are many ways of celebrating Thanksgiving in the world. American Thanksgiving Day is another important holiday related to Christianity. It is on the fourth Thursday of each November.

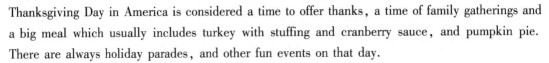

Thanksgiving Day in America is considered a time to offer thanks, a time of family gatherings and a big meal which usually includes turkey with stuffing and cranberry sauce, and pumpkin pie. There are always holiday parades, and other fun events on that day.

Exercise 2

My name is Li Xia. I'm fourteen years old. I come from a small village in Yunnan. I am a student of Class 1, Grade 1. My school life is very interesting. We learn Chinese, English, math, history, biology, geography, politics, physics and so on. I like P. E. the best. My teachers and classmates are very friendly. I like my school life very much.

Exercise 3

Thank you for your letter of the 16th of this month. We shall be glad to enter into business relations with your company. In compliance with your request, we are sending you, under separate cover, our latest catalogue and price list covering our range of exported products. Payment should be made by an irrevocable and confirmed Letter of Credit. Should you wish to place an order, please fax us.

Exercise 4

The main difference between Chinese and Western eating habits in restaurants is that unlike the West, where everyone has their own plate of food, in China the dishes are placed on the table and everybody shares. At home, westerners place the dishes on the table and, using serving spoons to put the different foods on their plate at the beginning of the meal, and then may help themselves to more when finished. If you are being treated by a Chinese host, you will be prepared a ton of food. Chinese are very proud of their culture of cuisine and will do their best to show their hospitality.

Exercise 1

Mid-Autumn Day is a traditional festival in China. Almost everyone likes to eat moon-cakes on that day. Most families will have a dinner together to celebrate the festival. A saying goes, "the moon in your hometown is almost always the brightest and roundest". Many people who live far away from their hometowns want to go back to have a family reunion. How happy it is to enjoy moon-cakes while watching the full moon with your family members.

Exercise 2

This year's summer vacation was most enjoyable. I spent fifteen days helping my grandparents doing farm work in the countryside, where I saw mountains and fields covered with green plants. Sometimes I went swimming in the river to the west of the village.

Every morning I helped the children in the neighbourhood with their lessons. All of them showed much interest in English. Their parents all thought highly of me. I now realize that knowledge is very much needed in the countryside.

Copy Version and Key to Each Exercise

Exercise 3

Things which were considered impolite many years ago are now acceptable. To see a woman smoking in public today is not surprising, nor is it always "ladies first" when going through doors, nor is it necessary for a man to give his seat to a woman on a crowded bus or train. Customs also differ from country to country. The important thing to remember about social customs is not to do anything that might make other people feel uncomfortable, especially if they are your guests.

Exercise 4

Dim sum, a kind of Chinese dish, is served from 6 a.m. in the morning to around 12 noon. It is mostly breads, and meats and vegetables wrapped in a kind of pastry. At a tea house where dim sum is served, the patrons sit around a large table and are served tea. Hot water is also provided to wash the plates, bowls, chopsticks and spoons.

Exercise 1

Different people choose different jobs for their ideal careers. As for me, I have made up my mind to be a teacher chiefly for three reasons. First of all, I want to teach because I like to teach students about what I have learned. Secondly, I like the freedom to do something for myself. Also, I have the opportunity to keep on learning.

Teaching is not an easy job at all. Therefore, I have to study hard to acquire more knowledge. And, at the same time, I will make every effort to purify my soul so that I can become a good example to others.

Exercise 2

Many thanks for your letter and enclosures of September 12th. We were very interested to hear that you are looking for a UK distributor for your teaching aids. We would like to invite you to visit our booth, No. 6, at next month's London Toy Fair, at Earl's Court, which starts on October 2nd. If you would like to set up an appointment during non-exhibition hall hours, please call me. I can then arrange for our staff to be present at the meeting. We look forward to hearing from you.

Exercise 3

When I first passed through the school gate and saw my new classrooms and teachers, I knew I was beginning a new stage in my life. School life is very busy and exciting. There are many activities in our school, such as The Sports Meet, Science Week and Art Week and so on. We all take an active part in them. I have made many new friends and we often help and learn from each other in many different respects. Our teachers have taught us a lot. They not only teach us how to learn, but also teach us how to be a helpful person in society.

Test

The Lantern Festival is a traditional Chinese festival. It is celebrated on the 15th day of the first month of the lunar year. The Lantern Festival is one of the biggest holidays in China. Several days before it, people begin to make lanterns. Lanterns are made in the shape of different animals, vegetables, fruits and many other things. While making lanterns, people usually write riddles on them. On the eve of the Lantern Festival, all the lanterns are hung up. On Lantern Festival people go outside to have a look at the lanterns and guess the riddles on them. Perhaps you can see some wonderful folk performances, Dragon Dance and Yangko. Everything is very interesting and so colorful.

Copy Version and Key to Each Exercise

Unit 4

Practice Exercise

Task

Exercise 1

M: It says in the newspaper that the weather today will be rainy and cold again.

W: Well, to tell you the truth, I don't like this kind of weather. I prefer it warm and dry.

M: I agree with you. Weather like this makes me feel tired. I hope it will be sunny this weekend, so I can go fishing with my friends.

W: Tom. What's the weather like in London in this season?

M: Well. Cold, of course, and sometimes it's snowy for a few days. How about the weather in Sydney now, Mary?

W: Pretty hot. Now it's summer there. I like the blue sky and the hot sea winds of summer in Australia.

M: Yes. That sounds great.

Key

Items	Information
Weather today	Rainy and cold.
Weather Mary likes	Warm and dry.
What is Tom going to do this weekend?	To go fishing with his friends.
Weather in London now	Cold and snowy.
Season in Sydney now	Summer.

Exercise 2

W: Dinner is ready.

M: Wow, Sally. I didn't know you are such a good cook. All the dishes look so delicious.

W: Thank you, Li Ming. Just help yourselves to anything you like.

M: It's a surprise for me that you know how to make Cantonese food.

W: I like Cantonese food. It's light and tastes good. How about you, Li Ming?

M: Well, I like it, too, but my favourite is Sichuan food.

W: Sichuan dishes are delicious, but they're too spicy.

M: Maybe you can try some Shanghai food. It's a bit heavier than Cantonese food, and it uses a lot of seafood and fish.

W: Oh, I'm fond of seafood.

M: Really. I know a famous restaurant which serves very good Shanghai food. Maybe we can try it this Sunday evening.

W: Great. I can't wait.

Key

Items	Information
Food Sally likes	Cantonese food.
Li Ming's favourite food	Sichuan food.
Food Li Ming suggests	Shanghai food.
Ingredients in Shanghai food	Seafood and fish.
Time to taste some new food	This Sunday evening.

Exercise 3

W: Good morning, Mr. Huang.

M: Good morning, you are Mary White, right?

W: Yes. Nice to meet you.

M: Well, please tell me something about your education first.

W: I'll graduate this June. My major is Business English.

M: What's your language ability?

W: I've passed PETS-4.

M: Good. Do you know how to use a computer?

W: Yes. I'm familiar with Microsoft Office software.

M: That's what we want. And what are your interests?

W: I like traveling very much.

M: That's all, Miss White. We'll inform you soon.

W: Thank you, Mr. Huang.

Key

Items	Information
Interviewee's name	Mary White.
Major	Business English.
Language ability	PETS-4.
Computer skill	Familiar with Microsoft Office software.
Hobbies	Traveling.

Copy Version and Key to Each Exercise

Exercise 4

W: Can I help you, sir?

M: Yes, I bought this MP3 player here the day before yesterday, but this morning I found it simply didn't work.

W: Let me have a look.

M: You see that you can't hear anything when you switch it on.

W: This is strange. We've been selling this kind of player for months and we haven't heard any complaints so far.

M: Well, I'm sorry, but I'm sure it is not my fault.

W: I'll change it for another one for you. Do you have the receipt?

M: Yes, here it is.

Key

Items	Information
What did the man buy?	MP3 player.
What's the problem?	It didn't work.
What did he hear when he switched the MP3 player on?	Nothing.
How will the woman solve the problem?	To change another one.
What does the woman ask for?	The receipt.

Exercise 5

M: Excuse me. Could you please tell me how to get to the Summer Palace?

W: Sure. Walk down this road; take the fourth turn to the right. Then you'll see it.

M: Is it far from here?

W: No. It's only about five minutes' walk.

M: OK, so I walk down this road; take the fourth turn to the right. It's only about five minutes' walk, right?

W: Yes, exactly. You can't miss it.

M: Many thanks!

W: Not at all.

Key

Items	Information
Destination	The Summer Palace.
How far is it?	About five minutes' walk.
Which turn?	Take the fourth turn.
To the right or left?	To the right.
Is it far?	No.

Exercise 6

W: Are you ready to order now, sir?

M: Yes. I would like the roast beef special.

W: You have a choice of vegetables: green peas or lima beans.

M: I'll have the green peas. And make sure the beef is well done.

W: Yes, sir. What would you like to drink? Coffee, tea or milk?

M: A cup of coffee, please, with cream and sugar.

W: The cream and sugar are on the table, sir.

M: Oh, yes.

W: Would you like to order some dessert?

M: I'll have fresh fruit and chocolate cake.

Key

Items	Information
What kind of beef	Roast.
Vegetable ordered	Green peas.
Drink ordered	A cup of coffee.
Things on the table	Cream and sugar.
Dessert ordered	Fresh fruit and chocolate cake.

Exercise 1

W: Good morning, Cambridge Theater.

M: Good morning. Have you got any tickets for the pop concert on September 14th?

W: Certainly, sir.

M: What time does the concert start?

W: At 8:00 p.m. How many tickets would you like?

M: Three, please.

W: What's your name?

M: Peter Brown.

W: Could I have your phone number, please, sir?

M: Of course, 7867254.

W: 7867254.

M: How much would that cost?

W: 64 dollars all together, sir.

M: Good.

W: We'll hold the tickets at the door until 7:30.

M: Thank you very much.

Copy Version and Key to Each Exercise

Exercise 2

M: Good evening. Is that the switchboard operator?

W: Yes, what can I do for you, sir?

M: Could you please give me a morning call at 5:40 tomorrow?

W: Sure. And your room number, please?

M: Room 1525, Jack Smith.

W: All right. An early call at 5:40, Room 1525, Jack Smith.

M: Please don't forget; otherwise, I will miss my appointment.

W: I won't. Have a nice sleep.

M: Thank you. Good night.

Exercise 3

W: Good morning, ABC Company. Can I help you?

M: I'm Tom Gates from Far-East Corporation. I'd like to make an appointment with Mr. Chen this week. I want to see him about some details of the contract.

W: Let me check Mr. Chen's diary. Well, would 9:30 Tuesday morning be convenient for you?

M: Yes. That'll be fine.

W: Then, next Tuesday morning, at 9:30, Mr. Chen will meet you in the office. And, Mr. Gates, would you mind leaving your contact number?

M: My number is 3546891.

W: Thank you, sir.

Exercise 4

W: Good morning, Mr. Walker's office.

M: Good morning. This is David Smith. I have an appointment with Mr. Walker at 10:00 tomorrow morning.

W: Yes, that's right, Mr. Smith.

M: I'm afraid I can't come then. I have to attend an important meeting in New York tomorrow. I will leave now and come back on Friday. Can we make it some other time next week?

W: Well, would 9:00 next Tuesday morning be convenient for you?

M: Yes, that'll be fine.

W: Then next Tuesday morning, at 9:00, Mr. Walker will meet you in his office.

M: Thanks a lot. Bye.

W: Bye.

Unit 5

Practice Exercise

Task

Exercise 1

A: Hi, David. How are you?

B: Fine. Yourself?

A: Good. Will you be free tomorrow night?

B: Yes, why?

A: We're having a birthday party for Miss Chen, and we'd be happy if you could come.

B: With pleasure. What time should I be there?

A: Oh, about 6:30 p.m. Are you free at that time?

B: No problem.

A: Shall I pick you up at your office at six?

B: I'll be waiting for you. Thank you.

A: Don't mention it. Bye.

B: See you tomorrow.

Exercise 2

A: Good morning, Mark. Come in and sit down, please.

B: Thanks.

A: Well, have you been in touch with Mr. Green recently?

B: Yes, I talked to him yesterday morning. He'll be back tomorrow.

A: Good. The meeting will be held the day after tomorrow. Could you be there, too?

B: Well... I've arranged to see my doctor because I had an accident last month.

A: What time are you seeing your doctor?

B: 2 o'clock.

A: That shouldn't cause any problems. We're going to have our meeting at 4 o'clock.

B: All right. I'll get back to my office as soon as possible.

A: Great.

Copy Version and Key to Each Exercise

Exercise 3

A: Hi, Cathy. How about going for lunch together?

B: OK. I was just thinking about having lunch with you.

A: Shall we have Chinese food or Western food?

B: It is up to you.

A: How about trying some Chinese food? There is a restaurant serving Chinese food nearby.

B: Great.

A: Do you often eat Chinese food?

B: Once in a while.

A: What is your favourite Chinese food?

B: It is hard to say. I like Sichuan spicy food a lot, and also Beijing Roast Duck.

A: Oh, that restaurant is famous for Beijing Roast Duck. Let's go and try it.

B: Sounds good.

Exercise 1

Wang: Hello! David, Paul.

David: Hi, Wang Hai. We were looking for you.

Wang: Why? What's the matter?

Paul: We're doing a project about China. May I ask you something about your culture?

Wang: Sure. Go ahead.

Paul: I read in a book that Chinese people give red envelopes to children and unmarried young adults on New Year's Day. What does that mean?

Wang: Oh, there's actually some money inside the red envelopes. They're New Year's gifts.

David: Why in red envelopes?

Wang: Well, red is popular in China. It symbolizes happiness.

David: Sounds good. I wish we had the same custom. Children would be excited to get them.

Wang: You bet! All children look forward to this exciting moment. Do you have a similar custom in America?

David: No, no customs about money. We just have parties for New Year's Day.

Wang: Your Christmas season and our Spring Festival have a lot in common, I think.

Paul: True. New Year is also one of the important festivals for Americans.

Wang: Now, does this information help for your project?

Paul & David: Yes, it does. Thanks very much.

Exercise 2

Staff A/B: Good morning, Mr. Lin. Welcome back. How was your trip?

Manager: Everything went smoothly except for the bad weather. It rained every day. Anyway, what has been going on here in the office?

Staff A: Let me look in the file... Oh yes, two important things have happened. First, we've had some changes in the department. George has been replaced by Billy Wang.

Manager: And what about George?

Staff A: I was just coming to him. When he heard about the change, he resigned.

Manager: Well, that is news. What about the other important thing?

Staff A: Well, that would be a bit of a surprise for you, too.

Manager: Oh?

Staff A: You've been selected to go to a course next year. You will be going to Singapore for two weeks.

Manager: Great! I've been waiting for that course since last year. Two weeks in Singapore. Better than Hong Kong, I think.

Staff B: Well, Mr. Lin. I have one more important thing.

Manager: I hope it is good news.

Staff B: It is. The big American deal came through last week. They agreed to sign that contract with us.

Manager: Really? Well done.

Staff B: But you missed the celebration party.

Manager: That's fine with me.

Exercise 3

A: Hello. Xiao Liang. I've heard you have a part-time job.

B: Yes. I'm working three nights a week at a supermarket.

C: I'm working part time, too.

A: Do they give you decent wages?

C: Not very good. But you know, it's hard times and not easy finding a job.

A: You are right. Business is slack everywhere due to the financial crisis.

B: And you? You said you had just been accepted by an insurance company, so you'll be earning some extra money, too.

A: Yeah, but... um, may I ask you a personal question?

B: It depends. What do you want to know?

A: Could you tell me what you do with the money you earn? Do you spend it immediately or do you save it?

B: Well, I spend some of it on books and I also save some for future use.

A: In a bank?

B: Yes, I've opened an account in a bank.

A: Which one?

B: The Construction Bank on campus.

A: Thanks, I will do that, too.

Copy Version and Key to Each Exercise

Exercise 1

A: Hi, Good morning, Zhang Yi. And how are you?

B: Very well, thank you. And you?

A: I'm fine. First of all, I think we should go around and meet some of the staff you'll be working with. Let's start with Ling Yun, your boss.

B: Okay.

C: Come in, please.

A: Good morning, Ling Yun. I'm just showing Zhang Yi around, doing the introductions before he starts working with you.

C: Good morning, Zhang Yi. It's nice to meet you.

B: It's nice to meet you, too.

A: Now, you'll be working with him next week. In the second part of this week you will be shown how everything's done, so that you know your way around.

C: Fine. So, I'll look forward to seeing you next week, Zhang Yi.

B: Yes, I'm looking forward to working with you, Mr. Ling.

(In the office)

A: Good morning, everyone. I really appreciate your taking the time to allow me to introduce a new member of our staff. First things first, I'd like to introduce you to our Office Manager, Mr. Huang. You'll be spending today and tomorrow with him.

D: Pleased to meet you, my name's Huang Hao.

B: Glad to meet you. I'm Zhang Yi.

A: Zhang Yi. Let me introduce you to Mr. Chang, this is Zhang Yi.

E: How do you do? My name's Chang Cheng.

F: I'm Jiang Yan.

A: Jiang Yan is Ling Yun's Personal Assistant. She'll show you how everything works before she leaves us at the end of the week and then you'll take her place.

B: Nice to meet you, Jiang Yan.

F: Glad to meet you, too. Now I am going to show you how to file the documents.

Exercise 2

A: Good day, everyone. I've worked out a tentative itinerary for these four days in Beijing. Don't hesitate to make any changes. Tomorrow, I guess you need a rest so as to recover a bit from the long journey.

B: Thank you. That is very considerate of you. Since our stay here is very short, an agreeable agenda will surely be essential.

A: Tomorrow afternoon, I think we'll have a meeting about business affairs. We'll reconsider the present contract since it is going to expire next month.

C: I agree, but I think the present contract works quite well. As for a new one, I propose we simply have to add more terms to make it more reasonable and beneficial to both of us.

A: I have no objection to your proposal. Then, we'll talk about our cooperation.

C: Yes, I think that's a good idea.

A: Fine, but what would you suggest we discuss about it first, then?

C: To my mind, the first thing to settle upon for our business negotiation is the easy matters, such as mode of investment, prices, and method of payment.

A: To reach an agreement, I'm afraid we'll need at least two rounds of negotiations if everything goes as smoothly as we expect. Let's make it the 12th and the 13th, shall we?

B: That's fine. Let's try our best to arrive at a good result.

A: For those two days, we'll have the discussion in the conference room in this hotel. Lunch will be served in the cafeteria. Any items we can't agree upon, we can compile memos for further discussion, and hopefully the second day will result in signing a new trade contract and a cooperation agreement.

D: That sounds fine.

A: The 15th is Sunday and we'd like to take you to the Great Wall and the Ming Tombs.

D: Thank you, Miss Zhang. It's very kind of you. I wouldn't dream of leaving Beijing without seeing the Great Wall. I'd like to see the pandas, too.

A: OK. We can stop at the zoo for a while on our way to the Great Wall. You might want to get some postcards, too.

D: That is a splendid idea. I'll certainly buy a few T-shirts and also a dozen postcards, just in case. I've got my wife's nieces in mind to buy gifts for.

A: By the way. Your flight is at ten o'clock, Monday morning. Is that OK for you?

B: Well, I don't think we have any other choice, it's fine.

C: That's an excellent itinerary.

D: You're really a thoughtful planner. Thanks a lot, Miss Zhang.

Test

A: We're from the Corner Store. Here is my business card.

B: Welcome to our company. My name's Lin, from the Sales Department. Nice to meet you.

A: Nice to meet you, too. I hear that you have some interesting products.

B: Thank you, yes. We import them from several different countries.

C: What about this one? What is it used for?

B: It's called a salad spinner; it's used for drying lettuce.

C: Are there any metal parts in it?

B: No, it's all made of plastic.

D: Where does it come from?

B: It's made in France.

Copy Version and Key to Each Exercise

A: And what about this one? What's this one used for?
B: It's used for making pasta, you know, noodles and things like that. It's made in China.
D: Do these kitchen gadgets sell well?
A: Yes, they do. You know a lot of people show great interest in cooking these days. Our products are carried by department stores and by a lot of kitchen ware stores, too.
D: Don't these gadgets go in and out of fashion pretty quickly?
B: Some of them do, but things like food processors have become standard items.
D: Do you carry a line of food processors as well?
B: Yes, we do. Here's a sample. It's made in Japan.
A: I'm interested in this kind of stuff because I own a food store.
B: Here, take one of our catalogues. All of our equipment is listed in it.

Unit 6

Part III

Practice Exercise

Task

Exercise 1

You are the monitor of Class Three, Grade One. Christmas Day is coming. The teacher asks you to write a notice to all the students in your class and put over his idea for an English Christmas evening party. He wants to hold the party in the meeting hall of our school from 7:00 to 10:00 p.m. on Christmas Eve and asks every group in the class to give at least two performances, such as singing English songs, performing English plays, telling jokes or stories in English and reading English poems.

Key

Dec. 21st, 2009

Notice

An English Christmas evening party is to be held in the meeting hall of our school from 7:00 to 10:00 p.m. on Christmas Eve. Each group has to give at least two performances, such as singing English songs, performing English plays, telling jokes or stories in English and reading English poems. Please get everything ready and be at the party on time.

Lily (Monitor)

Exercise 2

A: Good morning, Alex.

B: Good morning, Miss Zhao.

A: Alex, is everything ready for the sports meet this Saturday?

B: Oh, not really. The sports meet to be held this Saturday has to be put off because of the heavy rain these days.

A: Oh, really? What a pity!

B: It's really a pity. Do we need to come to school that day?

A: Yes. All the students are required to come to school on Saturday morning as usual.

B: OK. Will there be any classes in the afternoon?

Copy Version and Key to Each Exercise

A: Oh no, there will be no classes that afternoon.
B: I see, but shall we have the sports meet sometime later?
A: Of course! Weather permitting, the sports meet will be held next Saturday morning.
B: That's great! I will tell the students about it.
A: Alex, could you do me a favor?
B: Yes, it will be my pleasure!
A: Please ask the members of the school basketball team to come to the basketball court at 4:30 this Saturday afternoon.
B: Sure, but why?
A: There will be a basketball match between No. 1 Middle School and our school at 5:00.
B: Fantastic! I will write a notice. Don't worry!
A: Thank you very much.
B: You are welcome!

Key

Apr. 10th, 2009

Notice

The sports meet <u>which was planned for this Saturday has to be put off because of the heavy rain these days.</u> All the students are required to come to school on Saturday morning as usual, but there will be no classes that afternoon. Weather permitting, the sports meet will be held next Saturday morning. Members of the school basketball team must come to the basketball court at 4:30 this Saturday afternoon. There will be a basketball match between No. 1 Middle School and our school at 5:00.

Alex (Secretary of the Sports Union)

Exercise 3

You are Joyce White, secretary of the Head Office of ABC Company. Please write a notice to announce about a new colleague, Sally Smith, appointment as a Sales Manager in your company. The appointment was made at the meeting last Tuesday. Sally is going to take office from the first day of next month. She is an expert on global marketing and sales. Please get ready for the welcome party to be held this Sunday.

Key

March 8th, 2010

Notice

A new Sales Manager (Sally Smith) <u>is appointed to the Head Office of our company. The appointment was made at the meeting last Tuesday. Sally is going to take office from the first day of next month. She is an expert on global marketing and sales. Please get ready for the welcome party to be held this Sunday.</u>

Thank you for your cooperation.

Joyce White (Secretary of the Head Office)

Exercise 4

A: Good morning, Ken. It seems that you are in a hurry.

B: Yeah... Recently we received a large number of export orders. We have to deal with a lot of foreign letters, emails and phone calls every day.

A: Yeah... We are so busy with foreign orders these days. I think we need some special training to raise our language ability.

B: That's right. It's extremely urgent, I think.

A: This morning, I got a notice about a language training course.

B: Oh, that's great! What does the notice say?

A: It says that a foreign language training course is going to be organized by the Training Department. It will start on the 15th of March and will last 2 weeks.

B: It sounds wonderful. Who will have the chance to attend the course?

A: Those who will be dealing with the export orders must attend this course.

B: Fantastic! Good news. I will surely attend it! Could you tell me the time?

A: We are required to go to the meeting room at 8:00 a.m. on the 15th of March, and be on time. The training course will begin at 8:30 a.m.

B: Thank you very much.

A: You are welcome!

B: See you then.

A: See you.

Key

August 18th, 2009

Notice

Recently we received a large number of export orders. It's extremely urgent for the staff of our company to raise their language ability. A foreign language training course is going to be organized. It will start on the 15th of March and will last 2 weeks. Those who will be dealing with the export orders must attend this course. Please come to the meeting room at 8:00 a.m. on the 15th of March, on time. The training course will begin at 8:30 a.m.

Thanks for your cooperation.

Ken Wang (Training Manager)

Copy Version and Key to Each Exercise

Part V

Practice Exercise

Task
Exercise 1

You are Mr. White, the Sales Accountant. Recently, you have received complaints from a customer in Singapore that you had sent him the wrong model. This had never happened before. Mr. King, the manager is very concerned about this issue. Write a memo to all the staff and apologize for your mistake and state your solutions to the problem.

Key

Memo

To: All the Staff

From: Mr. White (Sales Accountant)

Date: April 15th, 2010

Subject: Complaints from a customer in Singapore

Cc: Mr. King (the manager)

I have checked my records and I found that Mr. Smith is correct in his complaint. I am afraid I was responsible for sending him the wrong model. I must apologize for making this mistake. I promise that I'll write immediately to him to apologize and send him the right model.

I hope this suggestion is satisfactory.

Exercise 2

A: Excuse me, Mr. Smith. Is it OK to talk to you for a second?

B: Sure, Bob. You look troubled today. What's the matter?

A: Mr. Smith... You know I have worked for the Sales Development as a salesman for six years.

B: Yes, you have. Are you satisfied with this position in our company?

A: Yeah... of course, I have been.

B: That's great. You are doing a good job in our company, and I must thank you.

A: Don't mention it. Mr. Smith... Mr. Smith, a friend of mine introduced me to BBS Company recently.

B: Oh? Are they offering you a job?!

A: Yes. It's really a good offer... and I have decided to accept the position offered by them.

B: Oh... OK! So you are telling me that you will quit your position here?

A: Yes... Sorry... Mr. Smith. They promised to give me a higher salary. I have a big family, and I need the money.

B: Oh, I understand. I wish you a bright future!

A: Thanks a lot, Mr. Smith. Thanks for all your help over the years.

B: You're welcome. I sincerely hope the best for you.

A: I hope the best for you, too.

B: Please send me a memo about resigning from your job as soon as possible.

A: OK.

Key

Memo

To: Mr. Smith

From: Bob

Date: May 6th, 2010

Subject: Resigning my position

I have worked for the Sales Development as a salesman for six years, and I have been satisfied with this position. However, a friend of mine introduced me to BBS Company, and I have decided to accept a job offered there because they promised to give me a higher salary. I therefore write this memo so as to resign from my job. I am giving the required two-week notice.

Copy Version and Key to Each Exercise

Unit 7

Practice Exercise

Task

Exercise 1

Scene（情景）

Your name is Mary. This Sunday evening, you would like to have a party at your home to celebrate your graduation. You will invite all of your classmates. Write a letter to invite one of your classmates, Anna, to the party. In the letter, tell her the party will begin at 7:00 p.m.

Key

Dear Anna,
I wonder if you can come to my party at my home this Sunday evening. I would like to celebrate our graduation. I am going to invite all our classmates to share this important moment together. The party will begin at 7:00 p.m. I hope you can come and join us.
Look forward to your favorable reply.

Yours faithfully,
Mary

Exercise 2

Scene（情景）

Your name is Wilson. You are the chairman of the Student Council in your school, Guangming Financial Vocational School. Now you would like to invite Professor Smith to give a speech about Western cultures to the students at your school, and show your great interest in it. Write a letter as an invitation.

Key

> Dear Professor Smith,
> I am the chairman of the Student Council in Guangming Financial Vocational School. <u>We would very much like to have you give a speech to our students. Many of us are interested in Western cultures. We hope you can share something about this with us.</u>
>
> Look forward to your favorable reply.
>
> Yours faithfully,
> Wilson

Exercise 3

Dialogue（对话）

A：Hi, Tom.

B：Hello, Lucy.

A：How's it going?

B：Pretty good. I like this new school, and everyone here is very kind.

A：I am glad that you like it here. Ah, Tom. We will have a dance party this Saturday evening. Would you like to join us? You can make more friends there.

B：That's great.

A：Can you ask Ellen to come with you?

B：Well, I think I will have to write an e-mail to her. We don't live in the same neighborhood.

A：I hope she can come.

B：She likes dancing. I think she will be happy to come.

A：Oh, I nearly forgot to tell you the time and place. 7：00 p.m., Saturday, at the Summer Club.

B：7：00 p.m., Saturday, at the Summer Club. Is that club's meeting room near the gate of our school?

A：Yes.

B：Well, maybe Ellen doesn't know how to get to our school. I think I will plan to meet her at her home at 6：30.

A：Great. See you then.

B：OK. Talk to you soon.

Copy Version and Key to Each Exercise

Key

Dear Ellen,

I'm writing to invite you to our dance party. The party will start at 7:00 p.m. on Saturday evening at the Summer Club meeting room near my school. We will be very glad if you can come. If you would like to join us, I will meet you at your home at 6:30 p.m.

I would love for you to attend, so please let me know your decision.

Yours faithfully,
Helen Liu

Exercise 1

Scene (情景)

You are Li Ping. You just returned home from a trip to visit Jane in her hometown, London. You think this trip was very interesting. She showed you around London and invited you to have dinner with her family. You had a very happy and unforgettable trip there, and you hope your friendship will continue. Write a letter to thank Jane and invite her to visit Beijing next summer holiday. You'll show her around your beautiful hometown.

Key

Dear Jane,

I am writing to thank you for your kindness. During my visit to London, it was wonderful that you showed me around the city and invited me to have dinner with your family. I had a very happy and unforgettable trip. I hope that our friendship will go on forever. I would like to invite you to visit Beijing next summer holiday so that I can have a chance to show you around my beautiful hometown.

With kind personal regards,

Faithfully yours,
Li Ping

Exercise 2

Scene (情景)

You are Helen Wang. You have passed PETS-4. You know Mr. Brown spent a lot of time helping you with your English so that you could make a lot of progress and pass the exam. Now you are going to write a letter to thank him for this.

Key

Dear Mr. Brown,
I am writing to tell you a piece of good news. I have passed PETS-4. Great thanks should be given to you. You spent a lot of time helping me with my English, so that I made a lot of progress.
Thank you very much.

Sincerely yours,
Helen Wang

Exercise 3

Dialogue（对话）

A: Hi, Wang Ling.

B: Hello, Joe.

A: I heard that you got first prize in the English Competition yesterday.

B: Yes.

A: Wow, you are so great. Congratulations.

B: Thank you. It's all because of Mr. Smith's help.

A: John Smith? He is a very nice teacher. I like him, too. He went back to New York last week.

B: Yes. He will come back next term. I can't wait to tell him about it.

A: Well, you can send him a letter of thanks.

B: Good idea, but I don't know how to write it.

A: Let me help you. First, tell him the good news.

B: Yes, I will begin the letter like this, "I am writing to tell you a piece of good news. I won first place in the English Competition held in the city yesterday."

A: And then, show your thanks to him.

B: "Because of your help with my English, I made rapid progress. As a result, I was able to make such an achievement. I owe my success to you. Thank you very much."

A: Let me see. Well, that's good.

B: Thank you, Joe.

A: You're welcome.

Copy Version and Key to Each Exercise

Key

Dear Mr. Smith,

I am writing to tell you a piece of good news. I won first prize in the English Competition held in the city yesterday. Because of your help with my English, I have made rapid progress. As a result, I was able to make such an achievement. I owe my success to you. Thank you very much.

Best wishes to you.

Yours faithfully,
Wang Ling

Exercise 1

Scene（情景）

You are Huang Wei, the Sales Manager of Ningbo Textile Co., Ltd. You have learned from alibaba.com that GCT Corporation is a leading importer of textiles in France. Write a letter to introduce your company as specializing in the export of textiles and show your interest in establishing business relations with them. With the letter you should also enclose a copy of the new products' catalogue.

Key

Dear Sirs,

We've learned from alibaba.com that you are a leading importer of textiles in France. We are now writing to you in the hope of establishing business relations with you.

We specialize in the export of textiles in China. Our products enjoy great popularity in the world market.

Please let us have your specific enquiry if you are interested in any of the items listed in the catalogue. We shall make offers promptly.

Look forward to your favorable reply.

Yours faithfully,
Huang Wei
Encl. As Stated

Exercise 2

Scene（情景）

You are a British importer. Write a letter to Zhejiang Leather Product Co., Ltd. in China, and show your interest in establishing business relations with them. You got their address from the Chinese Chamber of Commerce in London. You are interested in various kinds of leather products and would like to get the catalogues and quotations from their company.

Key

Dear Sirs,

We obtained your name and address from the Chinese Chamber of Commerce in London. We are writing this letter in hopes to establish trade relations with you.

We are an importer of leather products and interested in various kinds of leather products. We would like to receive your catalogues and quotations.

We are looking forward to receiving your early reply.

Yours faithfully,
Joanna Wood

Exercise 1

Scene（情景）

You are Wang Ming, from GCT Company, China. Your company specializes in selling bicycles. You are now writing a sales letter to introduce your products. Your products have been highly recommended by customers from 52 countries for their lower prices and good quality. You are going to give a special discount of 10 % on all catalogue prices this month.

Key

Dear Sirs,

We are a company that can provide you with good bicycles which have been highly recommended by customers from 52 countries for their lower prices and good quality. In order to further popularize these products, all the catalogue prices are subject to a special discount of 10 % during this month only.

I believe there will be a good chance of starting business and even establishing partnership between us. And if you have any questions or want more information, please feel free to contact me.

Yours faithfully,
Wang Ming
Sales Manager

Exercise 2

Scene（情景）

You are a sales representative from Nanjing Green Leaves Import & Export Co., Ltd.. You

Copy Version and Key to Each Exercise

have learned that there is a great demand for Christmas toys in the international market. You write a sales letter to introduce your product—PR-TG0903 Christmas doll. Your product is stuffed with 100% high quality polyester and meets the international safety standards. You can make the size according to the clients' requirement.

Key

Dear Sirs,

We have learned that there is a great demand for Christmas toys in your market. We are pleased to introduce our PR-TG0903 Christmas doll, which is stuffed with 100% high quality polyester and it meets the international safety standards. We can also make the size according to your requirement.

For the detailed information, please find the enclosed catalogue No. 123.

Your immediate reply would be highly appreciated.

Yours faithfully,
Nanjing Green Leaves Import & Export Co., Ltd.
Mary Li

Encl. As Stated

Exercise 1

Dialogue（对话）

A: Ann, I think I have to write a letter of apology to G&M Trading Company.

B: What's the matter?

A: They sent us a letter on September 10th ordering toys for pets.

B: Yes, that's a big order.

A: There are so many orders that we are now out of stock, and we have to wait another six weeks for the new supplies.

B: Well, you should write them a letter to explain that, and you should promise to contact them when our new stocks come in.

A: Can you teach me how to write it?

B: Sure. First of all, you should thank them for their order of September 10th.

A: I got it.

B: Then, explain the reason why we can't fill their order at this time, and say sorry about it.

A: We regret to tell you that, owing to a shortage of stock, we are unable to fill your order at this time. We don't expect further deliveries for at least another six weeks.

B: That's right.

A: Shall I tell them the exact time we expect the new stocks to come in?

B: No, you needn't. Just tell them we will contact them as soon as possible.

A: We will contact you as soon as our new stock comes in. Is that all?

B: Yes, that's done.

Key

Dear Sirs,
Thank you for your order of September 10th for pet toys. We regret to tell you that, owing to a shortage of stock, we are unable to fill your order at this time. We don't expect further deliveries for at least another six weeks. We will contact you as soon as our new stock comes in.
Thanks for your understanding.

Yours faithfully,
Helen Liu

Exercise 2

A: Wang Gang, how about the shipment for New Trade Company?

B: I think we can't make it.

A: What's the matter?

B: I tried to get in touch with the manufacturers after receiving the e-mail, but they didn't reply until yesterday. It was not possible for me to fix the date of shipment.

A: Well, you should write a letter to New Trade Company to explain this matter.

B: OK. And what should I say?

A: Err... After receiving your e-mail, we tried to get in touch with the manufacturers and ask them for prompt delivery. As a result of their delay in replying, we couldn't fix the date of shipment and therefore didn't reply to your letter last month, for which we express our deep regret to you.

B: And then?

A: Then, have you set the shipping date?

B: Yes, it has been decided that the goods can be shipped on the S. S. Changjiang on the 21st this month.

A: Good. You should tell them this in the letter.

B: The goods will be shipped on S. S. Changjiang on the 21st of this month.

A: That's all?

B: I think you should tell them we will call them when the shipment is completed, or they might not cooperate with us next time.

A: You are right. Thank you.

Copy Version and Key to Each Exercise

B: You are welcome.

Key

> Dear Sirs,
> Thank you for your e-mail of June 20th. After receiving your e-mail, we tried to get in touch with the manufacturers and press them for prompt delivery. As a result of their delay in replying, we couldn't fix the date of shipment and didn't reply to your letter last month, for which we express our deep regret to you; however, it is now possible to ship the goods on the S. S. Changjiang on the 21st of this month.
> As soon as the shipment is completed, we shall call you the shipping advice without delay.
>
> Yours faithfully,
> Wang Gang

Exercise 1

Dialogue (对话)

A: Mary, do you remember John Smith?

B: Yes, the Sales Manager of the ABC Trading Company.

A: He got a promotion to Deputy Managing Director recently.

B: Wow, how nice. Then you should write him a letter of congratulations.

A: I know, but I don't know what to write.

B: Well, first of all, you should congratulate him on his promotion.

A: And then?

B: We have had a close association with him over the past ten years; we all know how well he works.

A: Right. He is a hard-working man. I think he is qualified for this important post.

B: Just write what you think of. Like this, "Because of our close association with you over the past ten years, we know how well you are qualified for this important post. You earned the promotion through years of hard work."

A: "Because of our close association with you over the past ten years, we know how well you are qualified for this important post. You earned the promotion through years of hard work."

B: And in closing, give him your wishes.

A: OK. "Again, congratulations and best wishes for continued success."

B: Perfect.

A: Thank you, Mary.

B: You are welcome.

Key

> Dear Mr. Smith,
> It is a pleasure to congratulate you on your recent promotion to Deputy Managing Director of ABC Trading Company. Because of our close association with you over the past ten years, we know how well you are qualified for this important post. You earned the promotion through years of hard work.
> Again, congratulations and best wishes for continued success.
>
> Sincerely yours,
> Blake Lee

Exercise 2

Dialogue（对话）

A: Li Ming, have you read today's Guangzhou Daily?

B: Yes.

A: Mr. Brown has won the Gold Star Prize. He is a very capable and charming CEO.

B: It seems that you appreciate him very much.

A: Yes, and I would like to write a letter of congratulations to him, but I don't know what to write.

B: Let me help you. I think, maybe we should begin like this. "Dear Mr. Brown, please accept our warmest congratulations on your winning the Gold Star Prize."

A: OK, I got it.

B: Then you may sing high praise of his work.

A: He has devoted so much to his company. I think the achievements of his company and this award are very satisfying for him. His work has been recognized by others.

B: Well, how about this sentence? "You have long devoted yourself to the management of your company, and it must be of great satisfaction to you to have your work recognized."

A: "You have long devoted yourself to the management of your company, and it must be of great satisfaction to you to have your work recognized."

B: Then some wishes for him. That's all.

A: Umm... "May the future bring you even greater recognition of your noble service."

B: Yes, that's it.

A: Done. Thank you, Li Ming.

B: You are welcome.

Copy Version and Key to Each Exercise

Key

Dear Dr. Brown,

Please accept our warmest congratulations on your winning the Gold Star Prize. You have long devoted yourself to the management of your company, and it must be of great satisfaction to you to have your work recognized. May the future bring you even greater recognition of your noble service.

Best wishes!

Yours sincerely,
Li Yan

Exercise 1

Dialogue（对话）

A: Mary, have you got a reply from ABC Trading Company?

B: Not yet. The delivery time is coming due.

A: I think you had better write them a letter to urge them to make their delivery on time.

B: I think so. Can you teach me how to write it?

A: Sure. At the beginning of the letter, you should express our anxiety about the shipment. Note the contract number in the letter to get their attention.

B: Well, we are anxious to know about the shipment of our Order No. 123 for plastic toys.

A: Then remind them of the delivery time. Ask them to inform us of the delivery time as soon as possible.

B: As the contracted time of delivery is rapidly coming due, it is necessary that you inform us of the delivery time without any further delay.

A: Tell them the importance of the shipment arriving on time as we stated from the beginning.

B: In the beginning, we stated clearly the importance of shipping the goods on time.

A: At last, call their attention to delivery within the time stated.

B: We are in urgent need of these goods and insist on delivery by the time stated.

A: Well, that's all.

B: OK, let me send it to them now.

A: Remember to check for their reply every day.

B: I will.

Key

> Dear Sirs,
> We are anxious to know about the shipment of our Order No. 123 for plastic toys. As the contracted time of delivery is rapidly coming due, it is necessary that you inform us of the delivery time without any further delay. In the beginning, we stated clearly the importance of shipping the goods on time. We are in urgent need of these goods and insist on delivery within the time stated.
> We are looking forward to your early reply.
>
> Yours faithfully,
> Mary White

Exercise 2

Dialogue (对话)

A: Tom, how about the order No. 456? The date of the shipment should be in October.

B: I know, but I haven't received any information about the shipment yet.

A: Our end users phoned me again about the goods this morning.

B: Well, I am writing a letter to New Trading Corporation now.

A: Tell them our end users are in urgent need of the goods.

B: I have mentioned this in the letter.

A: Why not send our vessel "S. S. Fengqing" to Vancouver to pick up the goods?

B: That's a good idea. When can the "Fengqing" arrive there?

A: Around the end of October.

B: Let me give them this suggestion in the letter.

A: Ask them to give us a reply by fax immediately if they agree to this proposal.

B: OK.

A: If they don't accept this, ask them to inform us as to the earliest time the goods will be ready.

B: You are right. If not, please inform us of the earliest time when the goods will be ready.

A: Phone them if there is still no reply from them tomorrow.

B: Leave it to me.

Key

Dear Sirs,

Referring to the contract No. 456, the date of shipment is in October. However, up to now we have not received from you any information about the shipment. As our end users are in urgent need of the goods, we intend to send our vessel "S. S. Fengqing" to pick up the goods, which is expected to arrive at Vancouver around the end of October. Please let us have your immediate reply by fax if you agree to this proposal. If not, please inform us of the earliest time when the goods will be ready.

Awaiting your early reply.

Yours faithfully,
Tom Smith

Unit 8

Part II

Practice Exercise

Task 1

Exercise 1

Dialogue（对话）

Mrs. Li: I'm sorry that I called this meeting at such short notice, but it is urgent. Did everyone get the copy I handed out this morning?

Everybody: Yes, thank you.

Mrs. Li: That's very good. You can see from the copy that there are two matters we will discuss here today. First, what should we do to welcome the teachers from Peiying Middle School next Monday? Second, how are we going to celebrate Teachers' Day?

Mr. Wang: For the first item, I think we can ask the teachers from Peiying Middle School to attend some of our classes.

Mr. Liu: You are right. And after the classes, I think we can hold a meeting to share our teaching opinions.

Mrs. Li: I agree. An experience sharing meeting is important. Then, what about the Teachers' Day? Any ideas?

Mr. Liu: For many years, we have had big dinners to celebrate. This year, many teachers want to do something new.

Mr. Wang: Yes, the teachers in our office all suggested we go on a trip.

Mrs. Chen: Us, too.

Mrs. Li: Well, traveling seems a good idea, but, it takes much more money. We should discuss this with our headmaster. I'll declare the result next week. Many thanks to all of you.

Copy Version and Key to Each Exercise

Key

Meeting Minutes

Time: 3: 00 p.m., Friday, Sep. 2nd, 2009

Venue: Meeting room

Present: Li Zhaoji　　　Liu Haiyun

　　　　Wang Ming　　Chen Daming

Mrs. Li held the meeting to discuss the following two matters:

1. What to do to welcome the teachers from Peiying Middle school.
2. How to celebrate the Teachers' Day.

Suggestion for the first matter: To ask teachers to attend classes and have a meeting to share opinions.

Suggestion for the second matter: To go traveling.

Result will be declared: Next week.

Exercise 2

Dialogue (对话)

Headmaster: Now, let's start our meeting. Today I call you, all the teachers of the graduating classes, here to discuss a plan for the graduating students' practical training. You know, after this term, they are going to leave school to work. There are two months for their practical training. Let's see what ideas we can come up with to make this a good experience for our students.

Mrs. Gao: In my opinion, I think they have learned a lot about their major, but they don't have any practical experience. Going out into the workplace to put their skills into practice may be the best way. Do you think so?

Mrs. Zhang: Of course, practicing in a real workplace off campus is very important. I suggest that before they go out working, we should invite some successful students who have graduated from our school to share their work experience so that they can get ready for working in society.

Headmaster: I quite agree with both of you. What about other suggestions?

Mr. Zhao: I think students will be interested in some well known businessmen's stories. So, if possible, we can invite some successful businessmen to talk about how to be good employees and their own experiences.

Headmaster: Sounds good.

Mr. Zhao: A simulated job interview is important, too. It will help students get ready for the real interviews.

Headmaster: OK, now, we have four suggestions: first, working in a real workplace outside the college; second, inviting some graduating students to share their work experience; third, a successful businessman to give a speech; and fourth, a simulated job interview. That's good, thank you, all of you.

Key

Meeting Minutes

Time: 3:00 p.m., Tuesday, March 2nd, 2010
Venue: Meeting room
Present: Headmaster   Mrs. Gao
         Mrs. Zhang   Mr. Zhao
The meeting is held to discuss: a plan for the graduating students' practical training.
Suggestions in the meeting:
1. To work outside the college.
2. To invite some graduating students to share their work experience.
3. To invite some successful businessmen to give a speech.
4. A simulated job interview.

Exercise 1

Dialogue（对话）

Xiao Yan: Hello, everybody. Thank you for your time for this afternoon's meeting. Today, we are going to discuss our spring trip in early April. I call all the heads of every dormitory to share your ideas. Min Yi, will you please take down the notes?

Min Yi: No problem.

Xiao Yan: First, Zhang Xin, please tell us about your dormitory's opinion.

Zhang Xin: Well, I am from Room 305. All the members of our dormitory had a discussion last night. And we made our decision that we want to climb Baiyun Mountain.

Xiao Yan: What about your dormitory, Liang Hai?

Liang Hai: We, Room 306, also agree to go climbing mountains. We all want to get some fresh air.

Xiao Yan: Does Room 307 hold the same idea?

Li Jun: No, students in our dormitory all want to visit Yuexiu Park. They want to see the Five Goats, the symbol of Guangzhou.

Xiao Yan: Let's hear the opinion from Room 308. Hai Di, where do you want to go?

Hai Di: We have the same opinion with Room 305 and 306. We also like to climb mountains.

Xiao Yan: Then we can make our final choice now. Since three of the four dormitories prefer to climb mountains, then let's go to Baiyun Mountain, OK?

Everybody: OK.

Key

Copy Version and Key to Each Exercise

Meeting Minutes
Time: 11:00 a.m., Monday, March 8th, 2010
Venue: Classroom
Present: Xiao Yan   Min Yi   Zhang Xin
Liang Hai   Li Jun   Hai Di
Minutes taker: <u>Min Yi</u>
Item discussed in the meeting: <u>Spring trip in early April</u>
Opinion from Rooms 305, 306 and 308: <u>To climb Baiyun Mountain.</u>
Opinion from Room 307: <u>To visit Yuexiu Park.</u>
Result of the meeting: <u>To climb Baiyun Mountain.</u>

Exercise 2

Dialogue (对话)

Monitor: Should we start our meeting now?

Everybody: Yes.

Monitor: Today, I call all the leaders in our class to the first meeting of this term just to discuss the activities we are going to have this term. I hope everybody will use your imagination and share your ideas with us. First, Liu Jun, you are the leader in charge of P.E. class, do you have a suggestion?

Liu Jun: I think many students in our class like playing badminton. What about holding a badminton match?

Wang Xuan: I agree. Last term, our class had a football match and basketball match. This term, we can make a change.

Monitor: Do all of you agree to have a badminton match this term?

Others: Yes, it's good to have a change from last year.

Monitor: What other activities should we have? Please continue to speak your mind.

Yang Qiang: Many of the students in our class often complain it's hard to remember all the new English words. Why not plan an English Vocabulary Match so that all our classmates will pay more attention to studying English.

Three students: Good idea.

Monitor: I am also thinking of inviting some students who are good at studying to share their learning experience. What do you think of this?

Three students: I agree. It sounds like a great idea.

Monitor: OK, this term, we are going to have a badminton match, an English Vocabulary Match and an experience sharing meeting. Also, we will take part in the school singing match. So, we will have at least these four activities. Thank you. Let's stop here.

Key

Meeting Minutes

Time: 5:00 p.m., Thursday, March 4th, 2010

Venue: Classroom

Present: Monitor   Liu Jun

   Wang Xuan   Yang Qiang

   Other two leaders

Topic of the meeting: <u>To discuss activities for this term.</u>

Result of the meeting (four activities in the new term):

1. <u>A badminton match.</u>
2. <u>An English Vocabulary Match.</u>
3. <u>An experience sharing meeting.</u>
4. <u>The school singing match.</u>

Exercise 1

Dialogue（对话）

Mr. Smith: Today I call you here to talk about the invitation to speak at the conference in October.

David: I think it's a great opportunity.

John: I agree. It will be good for our company to be known more widely.

David: Yes, it's also good for the marketing of our products.

Mr. Smith: OK, you all think it's a good opportunity to take part in the conference. Then, let's go to the next question: Who is going to speak at the conference?

John: Well, it's an important event. I think you should give a speech yourself. It will show that our company thinks this conference is important.

David: That's right. Mr. Smith. You are the best person to give this speech.

Mr. Smith: OK, let's take some action now. John, please collect all the data related with the speech. And David, please write a report about our newest products. I am going to show our latest products in the conference so that we can have more orders soon.

David and John: OK, let's start the work.

Mrs. Smith: And let's meet again at 4:00 next Wednesday afternoon to have further discussion about this matter. Thank you for attending, John and David.

Copy Version and Key to Each Exercise

Key

Meeting Minutes
Time: 11:00 a.m., Monday, Feb. 8th, 2010
Venue: Meeting room
Present: <u>David, John and Mr. Smith</u>
Items discussed in the meeting:
<u>The invitation to speak at the conference in October.</u>
Points discussed and decisions made:
1. <u>Mr. Smith will give the speech.</u>
2. <u>John will collect data for speech.</u>
3. <u>David will write a report about newest products.</u>
4. Next meeting: <u>at 4:00 next Wednesday afternoon</u>

Exercise 2

Dialogue（对话）

Jack: I'm sorry to have to call this meeting at such short notice. Did you all get a copy of the sales figures?

Together: Yes.

Jack: Good. So you have seen from my memo the purpose of this meeting. First, we need to find out the reason for the drop in sales; and second, we should try to solve the problem.

Amy and Mary: OK.

Jack: Now, Amy, what do you think about this matter?

Amy: Well, there is a lot more competition out there now. Our prices seem a little high.

Jack: Well, I agree, but I don't think this is the most important reason. Mary, your idea?

Mary: In my opinion, the sales people are not very motivated. We need to do something to encourage them to get out there and sell.

Jack: That may be the important point. We need to do something to give them a push. What about the bonus system? How many sales people get bonuses now?

Dick: Not many.

Jack: Why?

Dick: It's really difficult to get a bonus. They have to make $60,000 in sales. That's a lot. Most people can only reach $40,000.

Jack: Then, it may work if we lower the sales quotas to $45,000. In this way, more sales people will work harder because $45,000 is much easier to get.

Dick and Mary: Yes, that is a good idea.

Key

Meeting Minutes

Time: 9:00 a.m., Monday, Feb. 8th, 2010

Venue: Meeting room

Present: Jack   Amy
         Mary   Dick

Two purposes of the meeting:

1. To find out the reason for the drop in sales.
2. To try to solve the problem.

Two reasons for the drop in sales:

1. Prices seem a little high.
2. The sales people are not very motivated.

Result: To lower the sales quotas (to $45,000) for a bonus.

Copy Version and Key to Each Exercise

Module 3

Task

Dialogue（对话）

Mr. Smith: Judy, I just got a phone call which said that our company's water supply will be turned off this morning.

Judy: Really? That's too bad. Many people will use the toilet during that time.

Mr. Smith: Yeah, that's what I am worrying about. I know we can use the toilet on the ground floor in the reception hall.

Judy: That's good, but how can we get there?

Mr. Smith: Oh, let me think... Yes, we can use the elevator or emergency stairs to gain access to the toilet.

Judy: OK, I've got it. So, when will the water supply be off?

Mr. Smith: Tomorrow, between 10:00 and 12:00 a.m.

Judy: OK, and we can use the toilet in the reception hall on the ground floor through the emergency stairs or the elevator.

Mr. Smith: That's it.

Judy: Right. I will put the notice on the board as quickly as I can.

Mr. Smith: Thanks a lot, Judy.

Judy: You're welcome.

Key

March 20th, 2010

Notice

Our water supply will be turned off tomorrow, between 10 and 12 a.m.

Staff may go to the toilet in the reception hall on the ground floor through the emergency stairs or the elevator.

I apologize for any inconvenience this may cause.

Judy Wang (Secretary)

Dialogue（对话）

Mr. Smith: I'm sorry that I call this meeting at such short notice, but it is very urgent.

Everybody: What is it?

Mr. Smith: This morning, I received a complaint letter from our company's main customer—Kremel's Industry. They said that they had received goods of faulty quality from us.

Johnny Brown: How can it be? Our quality is always first class.

Mr. Smith: Well, they're unsatisfied with the products. So, today, we are going to have a discussion about the quality of our products. First, are there any problems in our assembly line? If there are, we need to find solutions to the problems. Johnny, do you have any idea what the problems could be?

Johnny: Umm... There should not be any problems in our producing technology, but this month, we've just changed to another supplier for our raw materials in Shanghai. I think we should examine the new materials immediately.

Mr. Smith: Yes. That's right. Simon, please do that immediately!

Simon Edgar: OK. I'll write you a report on this matter as soon as possible.

Mr. Smith: Now, how can we solve the problem with Kremel's Industry?

Steven Lee: Usually, if it's really our fault, we can ask them to ship the goods back and we reproduce new ones for them.

Mr. Smith: Yes. If they don't agree, we should do our best to meet their requirements since this is our main customer.

Everyone: Yes.

Mr. Smith: Well, when the result of the examination comes in, I will write a letter to Kremel, this afternoon I hope. Many thanks to all of you.

Key

Meeting Minutes
Time: 10:00 a.m. to 10:40 a.m., March. 20th, 2010
Venue: Meeting Room 1
Present: Mr. Smith (General Manager)
Steven Lee (Sales Manager)
Johnny Brown (Production Manager)
Simon Edgar (Supervisor of R&D Department)
Susan Wang (Merchandiser)
Judy Wang (Secretary)
Mr. Smith held the meeting to discuss the following two matters:
1. Problems in our assembly
2. Solutions to solve the problems with Kremel's Industry
Results of the discussion:
1. Simon Edgar should examine the new materials and write a report after that.
2. To ask Kremel to ship the goods back and reproduce new ones for them.

Copy Version and Key to Each Exercise

Judy: Hello, this is Judy speaking.

Smith: Judy, this is Mr. Smith. Could you write a memo to all the staff in the Sales Department for me?

Judy: Yes, of course. What is it about?

Smith: The subject of the memo is "Newly-appointed Sales Manager".

Judy: OK. "Newly-appointed Sales Manager." Who is he or she?

Smith: Joyce Chen. This appointment was made at the board meeting last Tuesday and she is going to take office on the first day of next month.

Judy: Yes... Appointment last Tuesday... Take office the first day-next month. Anything more?

Smith: Yes. You should also write that she is an expert on global marketing and sales.

Judy: OK... Expert on global marketing and sales.

Smith: Finally, ask the staff to get ready for a welcome party to be held on Sunday.

Judy: Welcome party... on Sunday. Is that all?

Smith: Yes, thank you very much.

Judy: You are welcome.

Key

Memo
To: All the Staff in the Sales Department
From: Mr. Smith (General Manager)
Date: March 20th, 2010
Subject: Newly-appointed Sales Manager
I am pleased to announce that Joyce Chen is appointed as our new Sales Manager in our company. The appointment was made at the board meeting last Tuesday and she is going to take office on the first day of next month. She is an expert on global marketing and sales. Please get ready for a welcome party to be held on Sunday.   Thank you for your cooperation.

Module 4

Task

W: Wow, Jack. You have a big family.

M: Yes. And I love all of them.

W: Who's this man beside you?

M: He's my grandpa. He's retired now. He used to be a doctor.

W: And this woman should be your grandma.

M: Yes. She's a housewife. She's very kind.

W: How about your parents?

M: They're all English teachers.

W: Do you have brothers or sisters?

M: Yes. I have an elder brother and a younger sister. This is my brother, Jim. He's twenty-one. He's a college student now. And this one is my sister, Ann. She's thirteen years old. She studies in a middle school in New York.

W: Well, you certainly have a lovely-looking family.

Key

Items	Information
Job of Jack's grandpa	A retired doctor.
Job of Jack's grandma	A housewife.
Job of Jack's parents	English teachers.
Jim's age	Twenty-one.
Location of Ann's school	In New York.

M: It says in the newspaper that the weather today will be rainy and cold again.

W: Well, to tell you the truth, I don't like this kind of weather. I prefer it warm and dry.

M: I agree with you. Weather like this makes me feel tired. I hope it will be sunny this weekend, so I can go fishing with my friends.

W: Tom. What's the weather like in London in this season?

M: Well. Cold, of course, and sometimes it's snowy for a few days. How about the weather in Sydney now, Mary?

W: Pretty hot. Now it's summer there. I like the blue sky and the hot sea winds of summer in

Copy Version and Key to Each Exercise

Australia.

M: Yes. That sounds great.

Key

Items	Information
Weather today	Rainy and cold.
Weather Mary likes	Warm and dry.
What is Tom going to do this weekend?	To go fishing with his friends.
Weather in London now	Cold and snowy.
Season in Sydney now	Summer.

3

W: Hello, Bill. How are you these days?

M: Hi, Mary. Not so well recently. I always feel tired and sick.

W: What's the matter with you? You look a bit pale.

M: Yes. I have to make a plan for a new product. I think I must be over-worked.

W: Don't work yourself so hard. I think you'd better have a good rest soon.

M: Yes, I think so. I have slept more than 10 hours today, but I still feel sleepy.

W: Oh, no. Bill, I think you should do some exercises in your spare time. I have joined the Mountain-Climbing Club. I think climbing mountains is an interesting sport, and the fresh air is good for our health. Would you like to join us?

M: Well, I'll think about it. But, you know, I have no free time.

W: Come on, Bill. If you want to, you can find the time.

Key

Items	Information
How is Bill recently?	Not so well, tired and sick.
Bill's task recently	To make a plan for a new product.
Mary's advice	To do some exercises.
Club's name	Mountain-Climbing Club.
Why is climbing mountains good sport?	Because the fresh air is good for our health.

4

W: Dinner is ready.

M: Wow, Sally. I didn't know you are such a good cook. All the dishes look so delicious.

W: Thank you, Li Ming. Just help yourself to anything you like.

M: It's a surprise for me that you know how to make Cantonese food.

W: I like Cantonese food. It's light and tastes good. How about you, Li Ming?

M: Well, I like it, too, but my favorite is Sichuan food.

W: Sichuan dishes are delicious, but they're too spicy.

M: Maybe you can try some Shanghai food. It's a bit heavier than Cantonese food, and it uses a lot of seafood and fish.

W: Oh, I'm fond of seafood.

M: Really. I know a famous restaurant which serves very good Shanghai food. Maybe we can try it this Sunday evening.

W: Great. I can't wait.

Key

Items	Information
Food Sally likes	Cantonese food.
Li Ming's favorite food	Sichuan food.
Food Li Ming suggests	Shanghai food.
Ingredients in Shanghai food	Seafood and fish.
Time to taste some new food	This Sunday evening.

W: David, what did you watch on TV yesterday?

M: I watched some game shows. I like to try to answer the questions with the contestants.

W: How boring! I watched a movie on Channel Six. It was a sad love story. I was moved to tears.

M: Well, I don't like that kind of movie. I prefer the action movies.

W: Did you watch the live concert by Michael Jackson?

M: No. When was it?

W: Yesterday afternoon. I know you like him so much.

M: Oh, what a pity!

W: It doesn't matter. There will be a replay at 8:00 Tuesday evening on Channel Three.

M: Really? 8:00 Tuesday evening on Channel Three. I can't miss it this time.

Key

Items	Information
TV programs David watched	Game shows.
The movie Anna watched was about	A sad love story.
David's favorite movies	Action movies.
Whose live concert	Michael Jackson's.
Replay time	At 8:00 Tuesday evening.

Copy Version and Key to Each Exercise

W: Air China. Good morning. Can I help you?

M: Good morning. I'd like to book a ticket to London on January 10th.

W: Let me check. Well, sir, Flight 682 will leave for London at 10:00 a.m. that day. There are still some seats available.

M: I'll book one business class.

W: Your name, please?

M: John Chen. J-O-H-N C-H-E-N.

W: Now you have been booked. John Chen, one business class, Flight 682 to London on January 10th. The departure time is 10:00 a.m.

M: Thank you. What is the fare?

W: US$500.

Key

Items	Information
Flight No. & destination	Flight 682 to London.
For whom	John Chen.
Departure time	10:00 a.m. on January 10th.
Ticket type	Business class.
Fare	US $500.

W: Good morning, Mr. Wang.

M: Good morning, you are Mary White, right?

W: Yes. Nice to meet you.

M: Well, please tell me something about your education first.

W: I'll graduate this June. My major is Business English.

M: What's your language ability?

W: I've passed PETS-4.

M: Good. Do you know how to use a computer?

W: Yes. I'm familiar with Microsoft Office software.

M: That's what we want. And what are your interests?

W: I like traveling very much.

M: That's all, Miss White. We'll inform you soon.

W: Thank you, Mr. Wang.

Key

Items	Information
Interviewee's name	Mary White.
Major	Business English.
Language ability	PETS – 4.
Computer skill	Familiar with the Microsoft Office software.
Hobbies	Traveling.

W: Good morning, ABC Company. Can I help you?

M: I'm Tom Gates from Far-East Corporation. I'd like to make an appointment with Mr. Chen this week. I want to see him about some details of the contract.

W: Let me check Mr. Chen's diary. Well, would 9:30 Tuesday morning be convenient for you?

M: Yes. That'll be fine.

W: Then, next Tuesday morning, at 9:30, Mr. Chen will meet you in the office. And, Mr. Gates, would you mind leaving your contact number?

M: My number is 3546891.

W: Thank you, sir.

Key

Items	Information
Caller's name	Tom Gates.
Caller's number	3546891.
To whom	Mr. Chen.
Message	To see him about some details of the contract.
Date/time	At 9:30, next Tuesday morning.

W: Excuse me, officer. Can you help me?

M: Sure.

W: Can you tell me how to get to Chinatown?

M: Well, it's a long way from here. I think you'd better take the subway.

W: Is there a subway station near here?

Copy Version and Key to Each Exercise

M: Yes, the station is just over there near the Grand Theatre. You should take Subway One to Mayflower Park.

W: Take the number one subway and get off at Mayflower Park. And how much is the fare?

M: 30 cents.

W: Thank you.

M: You are welcome.

Key

Items	Information
Place she wants to go to	Chinatown.
Transportation she should take	Subway One.
Location of subway station	Near the Grand Theatre.
Station to get off	Mayflower Park.
Fare	30 cents.

10

W: Tom, when is your National Day?

M: We call it Independence Day. It's on July 4th every year.

W: What do Americans do to celebrate this important day?

M: On that day, there will be parades during the daytime and firework displays in the evening. How about yours, Lily?

W: We'll have a seven-day holiday. Every year before National Day, my school will hold an evening party, and students will prepare some programs. This year we will put on an English short play. Will you come to see our play next Wednesday evening?

M: That sounds great. I can't wait for it.

Key

Items	Information
Tom's National Day	On July 4th.
How to celebrate in America	Parades, fireworks display.
How long is Chinese National Day's holiday?	7 days.
Lily's program this year	(Put on) An English short play.
When is Lily's program?	Next Wednesday evening.

11

W: Good morning, Mr. Walker's office.

M: Good morning. This is David Smith. I have an appointment with Mr. Walker at 10:00 tomorrow morning.

W: Yes, that's right, Mr. Smith.

M: I'm afraid I can't come then. I have to attend an important meeting in New York tomorrow. I will leave now and come back on Friday. Can we make it some other time next week?

W: Well, would 9:00 next Tuesday morning be convenient for you?

M: Yes, that'll be fine.

W: Then next Tuesday morning, at 9:00, Mr. Walker will meet you in his office.

M: Thanks a lot. Bye.

W: Bye.

Key

Items	Information
From whom	David Smith.
To whom	Mr. Walker.
Original meeting time	At 10:00 tomorrow morning.
Reason for not coming	To attend an important meeting in New York.
New meeting time	At 9:00 next Tuesday morning.

12

W: Michael, would you do me a favor?

M: Sure, Mary. What can I do?

W: I have a bad cold. Mr. Chen has granted me two days' leave.

M: Take care of yourself then.

W: Could you please make 10 copies of the report for Mr. Chen.

M: No problem.

W: And tomorrow afternoon, Mr. Smith will come to meet Mr. Chen in the office. Please receive him for me.

M: Yes, just leave that to me.

W: If there are any phone calls, please answer them for me. If something unexpected happens, please call me at 13598740382.

M: I see. Don't worry! Everything will be all right.

W: Thank you so much.

Copy Version and Key to Each Exercise

Key

Items	Information
Colleague's name	Michael.
The number of copies needed	10 copies.
Visitor's name	Mr. Smith.
Daily office work	To answer phone calls.
Mary's number	13598740382.

13

W: Excuse me.
M: What can I do for you, madam?
W: I'm staying in Room 1605. I want to have a rest now, but the guests next door are making too much noise. I hope you can do something.
M: I'll send a clerk to handle this matter immediately.
W: Thank you. And can you give me some more clean towels?
M: Of course.
W: And there's no hot water in the thermos.
M: OK, madam. I'll bring in some clean towels together with some hot water. We can try to fix the hot water once you have had your rest. Is there anything else?
W: Nothing else. Thank you.
M: If there's anything else you need, you can dial 83762419. I'm always at your service.

Key

Items	Information
Room number	1605.
What does she complain?	The guests next door are making too much noise.
How to handle the problem	To send a clerk to handle this matter immediately.
Room service	(Bring in) Some clean towels and hot water.
Phone number	83762419.

14

W: Hello, Mr. Grant. Did you move into your new house last week?
M: Yes, we did.
W: How do you like it?
M: We like it very much. We have two children, you know. Our son is twelve, and our

daughter is seven. We have four bedrooms in the new house.

W: Is your wife happy?

M: Yes, she is. The new kitchen is beautiful. She likes to cook in it.

W: Do you have a garden?

M: Yes, there's a big garden in the back yard. We have lots of flowers there.

Key

Items	Information
Time for moving into the new house	Last week.
Number of children	Two (children).
Age of the daughter	Seven.
Location of the garden	In the back yard.
Things in the garden	Lots of flowers.

15

M: Did you hear that Bob is in hospital?

W: Oh, really? What's the matter with him?

M: He's got a very high temperature. Something may be wrong with his lungs.

W: How in the world did he get that?

M: He just came back from India. He must have got H1N1 while he was there.

W: That's really too bad.

M: Oh, yes, but luckily, he seems to be improving. The doctor says if he stays in the hospital for a few weeks, he should be able to get better.

Key

Items	Information
Who's in hospital?	Bob.
What's the matter with Bob?	He's got a very high temperature.
Where did he come back?	He came back from India.
Disease	H1N1.
Present condition	Improving/Getting better.

16

M: We're having a picnic in Beiyun Park at 3 o'clock tomorrow afternoon. Why don't you come with us, Louise?

W: I'd like to, but I think it's going to rain. The weatherman says it is.

Copy Version and Key to Each Exercise

M: I don't think he's right. It hasn't rained for a week and it isn't cloudy today, either.
W: But he's always correct in his weather forecast.
M: The temperature is 20℃ this afternoon. I'm sure it'll be fine for our picnic.

Key

Items	Information
What are they going to do?	To have a picnic.
Time for the activity	At 3 o'clock tomorrow afternoon.
Weatherman says	It's going to rain.
What's the weather like now?	It isn't cloudy /fine.
Temperature this afternoon	20℃.

17

M: Come in, please, Mary. Come over and sit here. Your annual report is well done.
W: Yes, thank you, but I received a lot of help from my workmates.
M: I know you're a capable person. The others couldn't have helped you much. They had their own work to do.
W: Thank you, but I don't always do a good job.
M: Take it easy. Everyone makes mistakes. However, continue to be more careful. You have a bright future ahead of you.
W: Thank you so much, Mr. Smith. I'll do my best.

Key

Items	Information
Clerk's name	Mary.
Boss's name	Mr. Smith.
Result of her annual report	Well done.
Boss's comment on the clerk	A capable person.
Boss's suggestion	Try to be more careful.

18

W: Hello!
M: Hello! Ms. Jones. This is Robin Williams, David's father speaking.
W: Oh, hi, Mr. Williams. How's David now?
M: Much better. Thank you very much for helping David with his school work.

W: Don't mention it. That's what teachers should do.

M: We'd like to drop by your home this evening, just to express our thanks.

W: Please don't do that, thank you. I'll feel happy just seeing David back in class.

M: It's all so good of you. The doctor says David should be able to go back to school in two days.

W: That's great. See you then.

M: See you.

Key

Items	Information
Name of David's father	(Robin) Williams.
Teacher's name	Ms. Jones.
David's present condition	Much better.
Father's purpose for calling	Showing his thanks.
Time for David to go back to school	In two days.

19

M: Good morning, Susan! Umm...

W: What's up, Andrew? You look down today.

M: I got a "C" in the English exam again. What do you think I should do?

W: Well, learning English is easy, but it needs plenty of practice.

M: I did practice a lot, and I do lots of reading comprehension exercises every day.

W: Well, I suggest that every day you do some reading aloud, and try to speak more. That'll help you improve your sense of the language, which is essential for English learning.

M: I see. I'll give it a go.

Key

Items	Information
How did Andrew look today?	Down.
Mark/grade in the exam	"C".
How hard is he learning English?	Easy.
What should he do if he wants to learn English well?	To practice.
What should he do to improve his sense of language?	He should do some reading aloud and try to speak more.

20

M: Nicole, you've been watching TV for three hours today. One more minute in front of the

Copy Version and Key to Each Exercise

TV set, I'll cancel our trip to Disneyland tomorrow.
W: Please, Dad! It's Saturday today. You said I could watch more TV at weekends.
M: But you should learn to protect your eyes, dear. By the way, from today on, no more reading while lying in bed.
W: What about your smoking two packages of cigarettes a day, Dad?
M: Oh, you've got me there. I promise I'll try to smoke less and less from now on.
W: Let's make that a deal.

Key

Items	Information
How long has Nicole been watching TV today?	Three hours.
What day is it today?	Saturday.
What should Nicole learn to protect?	Eyes.
How many packages of cigarettes does the father smoke a day?	Two packages of cigarettes.
Dad's promise	To smoke less and less.

21

M: I'm terribly sorry, Sarah. Yesterday when I was hurrying home, I dropped your book on the ground and got it wet and dirty, and you know, it was raining hard at the time.
W: My book, *The God Father*, you borrowed it from me last week?
M: Sorry. Yeah. What can I do about it?
W: Oh, that's too bad. It was my favorite.
M: I do apologize for this. I'll buy a new one for you.
W: Never mind, Daniel. These things happen all the time. It's nothing to get upset about.

Key

Items	Information
What happened to the book?	Dropped on the ground and got wet and dirty.
Weather at the time	Raining.
Name of the book	*The God Father.*
When was the book borrowed?	Last week.
How to compensate	To buy a new one for him.

22

M: Excuse me. Could you please tell me how to get to the Summer Palace?

W: Sure. Walk down this road; take the fourth turn to the right. Then you'll see it.

M: Is it far from here?

W: No. It's only about a five minutes' walk.

M: OK, so I walk down this road; take the fourth turn to the right. It's only about a five minutes' walk, right?

W: Yes, exactly. You can't miss it.

M: Many thanks!

W: Not at all.

Key

Items	Information
Destination	The Summer Palace.
How far is it?	About a five minutes' walk.
Which turn?	Take the fourth turn.
To the right or left?	To the right.
Is it far?	No.

23

W: Hi, what can I do for you?

M: I'd like to buy a dress.

W: All our dresses are in this section. What do you think of this one here? It's made of silk.

M: Hmm, it looks nice, but I'd like to have something warm for the winter.

W: Maybe you would like a heavy wool dress. How about this one?

M: I think that's what I want. How much is it?

W: It's seventy-five dollars plus tax.

M: It's a little expensive. Do you think it's possible to get a discount?

W: Hmm, since you like it so much, how about a 10 percent discount?

M: That's good. Thank you.

Key

Items	Information
What to buy	A dress.
Material wanted	Wool.
For which season	For the winter.
Price	Seventy-five dollars plus tax.
Discount	10 percent discount.

Copy Version and Key to Each Exercise

24

M: Now, what seems to be the trouble?

W: It's nothing serious. But I always have a headache, and I haven't slept properly for several weeks. I've also lost my appetite and my eyes are burning.

M: Mmm, you do look rather pale. Let me take your temperature first. Now, let me listen to your pulse. Mmm, there's something wrong.

W: Yes, I never seem to have any energy.

M: Just as I thought. You've caught a cold. You'd better stay in bed for two days. And don't stay up late working. Try to get more rest.

W: Thank you.

Key

Items	Information
Mrs. Wang's problem	(Has a) Headache.
How well has Mrs. Wang slept?	She hasn't slept properly.
What does the doctor do then?	He takes her temperature/listens to the pulse.
Has Mrs. Wang had a cold?	Yes.
Doctor's suggestion	Stay in bed for two days/Don't stay up late working/Get more rest.

25

W: Do you have any hobbies, Frank?

M: Yes, I'm fond of fishing and I'm very keen on making home movies. What about you?

W: I like taking photographs, but I haven't got a movie camera.

M: I take a lot of photographs, too, but I'm interested in the history of the cinema. So I really enjoy using a movie camera.

W: What other interests have you got? Do you collect anything?

M: Yes, I collect stamps and I've got quite a big collection of records and tapes, but I don't buy many now.

Key

Items	Information
Frank's fond of _____	Fishing.
Frank's keen on _____	Making home movies.
Green's hobby	Taking photographs.
Frank's interested in _____	The history of the cinema.
Who collects stamps?	Frank does.

26

M: Today I had my first English class.

W: How was it?

M: It was interesting. The teacher gave us three suggestions to help ourselves learn the language.

W: That's good. What are they?

M: First, stop talking anything except English, then learn many complete sentences by heart, finally, have American friends tell us how they say different things, like expressions, and always imitate them.

W: Do you think that's right?

M: I couldn't agree more. It is necessary for us to imitate the native speakers.

Key

Items	Information
What class?	English class.
How was it?	Interesting.
First suggestion	To stop talking anything except English /Talking in English.
Second suggestion	To learn many complete sentences by heart.
Who to imitate	Native speakers.

27

M: My name's Steve Stone. Is the room I reserved ready?

W: Just a moment, please. Yes, it's ready. It's Room 305 on the third floor. Here is your key.

M: Thank you. I'll be staying here for five days. How much do you charge for the room?

W: 120 U.S. dollars a day. The check-out time is 12:00 at noon.

M: All right. I'll pay when I check out.

W: Thank you.

Key

Items	Information
What's the customer's name?	Steve Stone.
What is the room number?	Room 305.
How many days?	Five days.
How much a day?	120 U.S. dollars a day.
What is the check-out time?	12:00 at noon.

Copy Version and Key to Each Exercise

28

M: What time does the Museum and Art Gallery open?
W: At 9:00 a.m.
M: How much is the admission fee?
W: About 50 U.S. dollars for each person for all of the museums and the art gallery.
M: How many museums and art galleries do you have here?
W: The main museums you can visit are the Palace Museum and the History Museum, and there is one art gallery—Taipei Art Gallery.
M: Thank you for the information.
W: You're welcome. Have a nice day!

Key

Items	Information
Open time	(At) 9:00 a.m.
How much?	About 50 U.S. dollars.
How many museums?	Three.
Names of the museums	The Palace Museum / History Museum.
Name of the gallery	Taipei Art Gallery.

29

W: Can I help you?
M: I wonder where I can buy a map of the city.
W: You can get one in a bookstore or a post office.
M: Where is the nearest post office, then?
W: It's a bit far. So you may need to take a bus. Take the No. 5 bus at the corner of Nineteenth Street, and get off at the third stop on Twenty-Fourth Street. You will see the post office across the street.
M: Hmm, that's quite convenient. Thank you.
W: That's OK.

Key

Items	Information
What to buy	A map of the city.
Where to buy	In a bookstore or a post office.
Number of the bus	No. 5.
Place to take a bus	At the corner of Nineteenth Street.
Place to get off	At the third stop on Twenty-Fourth Street.

30

W: Could you meet Sally at the airport instead of me, David? Yesterday I promised to meet her, but I won't be free tomorrow.

M: Well, don't worry. I'll meet her for you. I guess I've met Sally before. Is she the short girl with short hair I saw at your birthday party last month?

W: Yes, and she remembers you very well.

M: Does she? When will her plane arrive?

W: Her plane will arrive at 4:25 p.m. from New York. The flight number is AA175.

M: AA175 from New York. I've got it.

Key

Items	Information
Who will meet Sally?	David.
Sally's appearance	A short girl with short hair.
Time for Sally's arrival	At 4:25 p.m.
Sally's coming from	New York.
Sally's flight number	AA175.

31

W: Good morning, Cambridge Theater.

M: Good morning. Have you got any tickets for the pop concert on September 14th?

W: Certainly, sir.

M: What time does the concert start?

W: At 8:00 p.m. How many tickets would you like?

M: Three, please.

W: What's your name?

M: Peter Brown.

W: Could I have your phone number, please, sir?

M: Of course, 7867254.

W: 7867254.

M: How much would that cost?

W: 64 dollars all together, sir.

M: Good.

W: We'll hold the tickets at the door until 7:30.

M: Thank you very much.

Copy Version and Key to Each Exercise

Key

Items	Information
Purpose of calling	For booking tickets for the pop concert.
Date for the concert	On September 14th.
Number of the tickets	Three.
Telephone number	7867254.
Where and when to get the tickets	At the door until 7:30.

32

W: May I see your passport?
M: Here it is.
W: How long are you going to stay?
M: For one week.
W: What is the purpose of your visit?
M: For sightseeing.
W: Do you have anything to declare?
M: No, nothing.
W: Do you have your air ticket and baggage claim tag?
M: Here they are.
W: Thank you, and would you fill in this AIR form, and also we need your signature here.
M: OK.

Key

Items	Information
What to see	Passport.
How long will he stay?	For one week.
Purpose for visit	For sightseeing.
Things to declare	Nothing.
What kind of form?	AIR form.

33

W: Are you ready to order now, sir?
M: Yes. I'll have the roast beef special.
W: You have a choice of vegetables: green peas or lima beans?
M: I'll have the green peas, and make sure the beef is well done.

163

W: Yes, sir. Would you like something to drink? Coffee, tea or milk?

M: A cup of coffee, please, with cream and sugar.

W: The cream and sugar are on the table, sir.

M: Oh, yes.

W: Would you like to order some dessert?

M: I'll have fresh fruit and chocolate cake.

Key

Items	Information
Kind of beef	Roast.
Vegetable ordered	Green peas.
Drink ordered	A cup of coffee.
Things on the table	Cream and sugar.
Dessert ordered	Fresh fruit and chocolate cake.

34

M: Yes, madam. Can I help you?

W: I hope so. I'd like to ask for some information about flights from Beijing to London.

M: Do you want to fly direct, or stop over somewhere on the way?

W: Oh, no, I want to go direct.

M: In that case, you have a choice of two airlines. C. A. A. C. flies once a week. It costs about 2,000 pounds for a return ticket and half of that for a one-way fare.

W: What is the other airline?

M: British Airways. It flies twice a week. 1,200 pounds for a one-way fare and 1,800 for a return ticket...

Key

Items	Information
Purpose of asking	Asking for information about flights.
The way of flying	Go direct.
How many airlines to choose from	Two.
Price of C. A. A. C. 's return ticket	About 2,000 pounds.
Price of B. A. 's one-way ticket	1,200 pounds.

35

M: Mary, is this your new computer?

Copy Version and Key to Each Exercise

W: Yes, I bought it last week.

M: How much was it?

W: 832 dollars.

M: It looks great.

W: Yes, it does. But...

M: What's wrong? You look worried.

W: Something's wrong with the Internet. I can't log on to the Internet.

M: Oh, let me check.

W: Thank you!

M: Oh... There's nothing wrong. Let me show you. It's easy.

W: OK.

M: Enter your password and address here. Look, it's OK!

W: I see. Thanks a lot.

Key

Items	Information
Owner of the computer	Mary.
When was the computer bought?	Last week.
Price of the computer	832 dollars.
Problem with the computer	Can't log on to the Internet.
Solution to the problem	Enter the password and address.

36

W: I have surprising news for you.

M: What's that, Alice?

W: Barbara is getting married.

M: Well, My best wishes to her. Who's the lucky guy?

W: Jason.

M: How romantic! When is the big day?

W: This August.

M: Good.

W: She asked if I'll be her maid of honor.

M: Did you promise her?

W: Yes, I did. It'll be my first time.

M: Who did Jason ask to be his best man?

W: My boyfriend, Tom.

M: That's very interesting!

Key

Items	Information
What's the surprising news?	Barbara is getting married.
Lucky guy's name	Jason.
Time to get married	This August.
Maid's name	Alice.
Best man's name	Tom.

37

M: Alice, do you have any plans for this holiday?
W: Oh, yes. I'm planning on going to Australia.
M: I've never been there. If I have a chance, I'll go there sometime.
W: Well. Would you like to go with us?
M: Great! When in the summer do you plan to leave?
W: It's best to go in mid-August, so... maybe on August 17th.
M: How long will it take to get there?
W: It'll take eighteen hours by air.
M: What's the weather like there?
W: It's hot.
M: What kind of clothes should I pack, then?
W: You'd better take some light cotton clothes.

Key

Items	Information
Where to go	Australia.
When to set out	Fairly early, on August 17th.
How long to get there	Eighteen hours.
How to get there	By air.
What kind of clothes to take	Some light cotton clothes.

38

W: Hi, Bob. Haven't seen you for a long time. How's everything going?
M: Not so well. I'm thinking of quitting my job.
W: Oh, I'm sorry to hear that. What kind of job would you like to do then?
M: I prefer a job full of challenges. I don't like to stay at the office all day long.

Copy Version and Key to Each Exercise

W: That's why you don't like to be a secretary, right?
M: You bet. I'd like to meet different people in different places every day.
W: Like what?
M: Salesman and tourist guide.
W: I hope you will succeed.
M: Thank you.

Key

Items	Information
Bob's present job	A secretary.
Bob is thinking of _____	Quitting his job.
What kind of job does Bob like?	A job full of challenges.
Bob's dislikes	Staying at the office all day long.
Bob's ideal jobs	Salesman and tourist guide.

39

M: Good afternoon.
W: Good afternoon. Take a seat, please.
M: Thank you.
W: We have received your application. I'd like to go over the details.
M: Yes, please.
W: I see that you're twenty-two and graduated from Business Administration University last July.
M: That's right.
W: What about your English?
M: I passed the Test for English Majors, Level Four.
W: By the way, what salary do you expect?
M: I really can't say. I have no experience and have to learn a lot.
W: Very good. We'll let you know the result in a week.
M: Thank you. Bye.

Key

Items	Information
Interviewee's age	Twenty-two.
University that Jack graduated from	Business Administration University.
Interviewee's English level	Passed the Test for English Majors, Level Four.
Expected salary	Really can't say.
Time to get the result	In a week.

40

W: Hey, Peter, what do you hope to do after graduating from Shenzhen Vocational School?

M: I'd like to work for a computer company. How about you, Mary?

W: My parents advised me to further my study at college, but I want to earn my own bread.

M: So it seems that you already know what you want.

W: I think it's very difficult for me to choose. I want to start my career soon to prove my ability; on the other hand, I think I will regret it if I miss going to college.

M: Well, whatever you do, I hope you can do it well.

W: Thanks. The same to you.

Key

Items	Information
What does Peter want to do after graduation?	To work for a computer company.
What's Mary's parents' advice?	Further her study at college.
What does Mary want to do?	To earn her own bread.
Mary will regret	If she misses going to college.
Peter's hope for Mary	She can do it well.

41

W: Excuse me, but are you Mr. Johnson?

M: Yes, I'm Mike Johnson from London.

W: How do you do? My name is Yang Hong. I'm from the S&K Import Company.

M: How do you do, Ms. Yang?

W: Welcome to Beijing, Mr. Johnson. Our manager asked me to come and meet you. Hope you had a good flight.

M: Thank you, Ms. Yang. I've had a very good flight.

W: Good. Let's take a short rest in the waiting room, and then we'll go to the hotel.

M: OK.

Key

Items	Information
Foreigner's name	Mike Johnson.
Name of the woman's company	S&K Import Company.
Where are they?	At the airport in Beijing.
Where to take a rest	A waiting room.
Where to go afterwards	The hotel.

Copy Version and Key to Each Exercise

42

M: Good evening. Is that the switchboard operator?
W: Yes, what can I do for you, sir?
M: Could you please give me a morning call at 5:40 tomorrow?
W: Sure. And your room number, please?
M: Room 1525, Jack Smith.
W: All right, an early call at 5:40, Room 1525, Jack Smith.
M: Right. Please don't forget; otherwise, I will miss my appointment.
W: I won't. Have a nice sleep.
M: Thank you. Good night.

Key

Items	Information
Purpose of calling	Having a morning call.
Time the call is wanted	5:40 tomorrow morning.
Caller's room number	Room 1525.
Caller's name	Jack Smith.
Reason for getting up early	Having an appointment.

43

W: Can I help you, sir?
M: Yes, I bought this MP3 player here the day before yesterday. This morning I found it simply didn't work.
W: Let me have a look.
M: You see you can't hear anything when you switch it on.
W: This is strange. We've been selling this kind of MP3 player for months and we haven't heard any complaints so far.
M: Well, I'm sorry, but I'm sure it is not my fault.
W: I'll change it for another one for you. Do you have the receipt?
M: Yes, here it is.

Key

Items	Information
What did the man buy?	A MP3 player.
What's the problem?	It didn't work.
What did he hear when he switched the MP3 player on?	Nothing.
How will the woman solve the problem?	To change another one.
What did the woman ask for?	The receipt.

44

M: Hello, Lee speaking.

W: Good morning, Mr. Lee. This is Alice.

M: Good morning, Alice.

W: I've just got a fax from my head office about the agency agreement. Would it be possible for us to meet sometime this afternoon?

M: I'm afraid I won't be available this afternoon. How about ten o'clock tomorrow morning?

W: Well, let me see... Oh, yes, I think I can manage that.

M: Good. I'll expect you then at 10:00 tomorrow morning, in my office.

W: OK. See you then. Bye.

Key

Items	Information
Caller's name	Alice.
Purpose of calling	To meet Lee.
Will Lee be free in the afternoon?	No.
When will they meet?	At 10:00 tomorrow morning.
Where to meet	At Lee's office.

45

W: Mike, have you ever bought anything on the Internet?

M: Yes, I have bought a handbag and two books.

W: Why don't you buy them in stores?

M: You know... the price is a bit lower than that at the local department store.

W: Sounds interesting. I have never tried that before. I want to buy some toys for Jack, my little brother, but I still have some questions.

M: Can I help you?

Copy Version and Key to Each Exercise

W: Yes, maybe. First, I don't know the website address for Internet shopping. Second, how do I pay online?

M: Don't worry. Let's surf the Internet and I'll tell you.

Key

Items	Information
Things bought on net	A handbag and two books.
Reason for buying online	A bit lower price.
What does the woman want to buy?	Some toys.
Who is Jack?	Woman's little brother.
Woman's second question	How to pay online.